Computational Rheology for Pipeline and Annular Flow

Computational Rheology for Pipeline and Annular Flow

by Wilson C. Chin, Ph.D.

Gulf Professional Publishing
an imprint of Butterworth-Heinemann

Boston Oxford Auckland Johannesburg Melbourne New Delhi

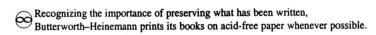

Library of Congress Cataloging-in-Publication Data
Chin, Wilson C.
    Computational rheology for pipline and annular flow/ by Wilson C. Chin.
    p. cm. Includes index
    ISBN 0-88415-320-7 (alk. Paper)
    1. Oil well drilling 2. Petroleum pipleines--fluid dynamics--mathematical models
    3. Wells--fluid dynamics--mathematical models. I. Title.

TN871.C49535 2000
622'.3382--dc21   00-06176

British Library Cataloguing-in-Publication Data
A catalogue record for this book is available from the British Library.

- Dedication -

To:

Mark A. Proett,
Friend and Colleague

Contents

Preface xi

1. **Introduction: Basic Principles and Applications** 1
    Why Study Rheology?, 5
    Review of Analytical Results, 7
    Overview of Annular Flow, 11
    Review of Prior Annular Models, 13
    The New Computational Models, 14
    Practical Applications, 17
    Philosophy of Numerical Modeling, 20
    References, 22

2. **Eccentric, Nonrotating, Annular Flow** 25
    Theory and Mathematical Formulation, 26
    Boundary Conforming Grid Generation, 33
    Numerical Finite Difference Solution, 35
    Detailed Calculated Results, 37
        Example 1. Fully Concentric Annular Flow, 37
        Example 2. Concentric Pipe and Borehole in the
                     Presence of a Cuttings Bed, 43
        Example 3. Highly Eccentric Circular Pipe and
                     Borehole, 48
        Example 4. Square Drill Collar in a Circular Hole, 50
        Example 5. Small Hole, Bingham Plastic Model, 52
        Example 6. Small Hole, Power Law Fluid, 57
        Example 7. Large Hole, Bingham Plastic, 59
        Example 8. Large Hole, Power Law Fluid, 61
        Example 9. Large Hole with Cuttings Bed, 65
    References, 68

3. Concentric, Rotating, Annular Flow **69**
General Governing Equations, 69
Exact Solution for Newtonian Flows, 72
Narrow Annulus Power Law Solution, 76
Analytical Validation, 80
Differences Between Newtonian and Power Law Flows, 81
More Applications Formulas, 83
Detailed Calculated Results, 85
    Example 1. East Greenbriar No. 2, 86
    Example 2. Detailed Spatial Properties Versus "r", 87
    Example 3. More of East Greenbriar, 96
References, 99

4. Recirculating Annular Vortex Flows **100**
What Are Recirculating Vortex Flows?, 101
Motivating Ideas and Controlling Variables, 102
Detailed Calculated Results, 103
How to Avoid Stagnant Bubbles, 110
A Practical Example, 111
References, 112

5. Applications to Drilling and Production **113**
Cuttings Transport in Deviated Wells, 115
    Discussion 1. Water-Base Muds, 115
    Discussion 2. Cuttings Transport Database, 120
    Discussion 3. Invert Emulsions Versus "All Oil" Muds, 122
    Discussion 4. Effect of Cuttings Bed Thickness, 125
    Discussion 5. Why $45° – 60°$ Inclinations Are Worst, 128
    Discussion 6. Key Issues in Cuttings Transport, 129
Evaluation of Spotting Fluids for Stuck Pipe, 131
Cementing Applications, 135
    Example 1.   Eccentric Nonrotating Flow, Baseline
                 Concentric Case, 136
    Example 2.   Eccentric Nonrotating Flow, Eccentric
                 Circular Case, 139
    Example 3.   Eccentric Nonrotating Flow, A Severe
                 Washout, 141
    Example 4.   Eccentric Nonrotating Flow, Casing
                 with Centralizers, 142
    Example 5.   Concentric Rotating Flows,
                 Stationary Baseline, 146
    Example 6.   Concentric Rotating Flows,
                 Rotating Casing, 150
Coiled Tubing Return Flows, 153

Heavily Clogged Stuck Pipe, 153
Conclusions, 155
References, 156

6. **Bundled Pipelines: Coupled Annular Velocity
    and Temperature** **159**
    Computer Visualization and Speech Synthesis, 160
    Coupled Velocity and Temperature Fields, 163
    References, 167

7. **Pipe Flow Modeling in General Ducts** **168**
    Newtonian Flow in Circular Pipes, 169
    Finite Difference Method, 170
    Newtonian Flow in Rectangular Ducts, 174
        Exact Analytical Series Solution, 174
        Finite Difference Solution, 177
        Example Calculation, 179
    General Boundary Conforming Grid Systems, 180
        Recapitulation, 183
        Two Example Calculations, 183
    Clogged Annulus and Stuck Pipe Modeling, 186
    References, 189

8. **Solids Deposition Modeling** **190**
    Mudcake Buildup on Porous Rock, 191
    Deposition Mechanics, 195
    Sedimentary Transport, 195
    Slurry Transport, 196
    Waxes and Paraffins, Basic Ideas, 197
    Hydrate Control, 201
    Modeling Concepts and Integration, 203
    Wax Buildup Due to Temperature Differences, 203
    Deposition and Flowfield Interaction, 204
    Detailed Calculated Examples, 205
        Simulation 1. Wax Deposition with Newtonian
                    Flow in Circular Duct, 205
        Simulation 2. Hydrate Plug with Newtonian Flow
                    in Circular Duct (Velocity Field), 210
        Simulation 3. Hydrate Plug with Newtonian Flow
                    in Circular Duct (Viscous Stress Field), 213
        Simulation 4. Hydrate Plug with Power Law Flow
                    in Circular Duct, 215
        Simulation 5. Hydrate Plug, Herschel-Bulkley Flow
                    in Circular Duct, 218

Simulation 6. Eroding a Clogged Bed, 222
References, 228

9. Pipe Bends, Secondary Flows, Fluid Heterogeneities 231
Modeling Non-Newtonian Duct Flow in Pipe Bends, 232
Straight, Closed Ducts, 232
Hagen-Poiseuille Flow Between Planes, 233
Flow Between Concentric Plates, 233
Flows in Closed Curved Ducts, 236
Fluid Heterogeneities and Secondary Flows, 238
References, 240

10. Advanced Modeling Methods 241
Complicated Problem Domains, 241
    Singly-Connected Regions, 242
    Doubly-Connected Regions, 243
    Triply-Connected Regions, 245
Convergence Acceleration, 246
Fast Solutions to Laplace's Equation, 247
Special Rheological Models, 249
Software Notes, 252
References, 253

Index 255

Author Biography 257

Preface

It has been a decade since *Borehole Flow Modeling* first appeared, integrating modern finite difference methods with advanced techniques in curvilinear grid generation, and applying powerful computational algorithms to annular flow problems encountered in drilling and producing horizontal and deviated wells. The early work combined essential elements of mathematics and numerical analysis, also paying careful attention to empirical results obtained from field and laboratory observation, and importantly, always driven by a strong emphasis on "real world" operational problems.

Prior to *BFM*, workers invoked "slot flow," "parallel plate," and "narrow annulus" assumptions to model non-Newtonian flows in eccentric annuli, with little success in correlating experimental data. With the new methods, which solved the complete flow equations exactly, at least to the extent permitted by numerical discretization, it was possible to *explain* the University of Tulsa's detailed cuttings transport database in terms of simple physical principles.

Fast forward to the new millennium. Cuttings transport, stuck pipe, and annular flow are even more important because deep subsea drilling imposes tighter demands on safety and efficiency. Rheology is more important than ever since drilling fluid characteristics now depend on temperature and pressure. To make matters worse, the same subsea applications introduce operational problems with severe economical consequences on the "delivery" side. More often than not, thick waxes will deposit unevenly and large hydrate plugs will form in mile-long pipelines. The harsh ocean environment and the lack of accessibility make every flow stoppage event a very serious matter involving millions of dollars.

While current industry interests focus on the thermodynamics of wax and hydrate crystallization and formation, and to some extent, on the altered properties of affected crudes, new rheological models will not be useful until they can be used with simulators to study macroscopic flow processes. For example, knowing "n" and "k" does not give the flow rate associated with a known pressure drop when large hydrate plugs or crescent-shaped wax deposits impede flow within the pipeline. This book uses the methods of *BFM* to model non-Newtonian flow in ducts with arbitrary geometrical cross-sections, e.g., different classes of blockages associated with different modes of wax deposition, hydrate formation, and debris accumulation.

xi

As in *BFM*, our advanced curvilinear grids adapt exactly to the cross-sectional geometry, and allow highly accurate numerical models to be formulated and solved ... *in seconds*. I will focus on practical results. For example, we will show that it is not unusual for "25% area blockages" to reduce volumetric flow by 60% to 80% or more. However, our new approaches provide more than "flow rate versus pressure drop" relationships. We will take advantage of new simulation capabilities to design dynamic, time-dependent models that show how wax, hydrates, and debris grow or erode with continuously changing duct cross-sections, where these changes are imposed by the velocity and stress environment of general non-Newtonian fluids.

Other applications are possible. For example, natural gas hydrates may become an economically viable energy source if potential logistical problems can be addressed satisfactorily. Some have suggested mixing ground hydrates with refrigerated crudes to form transportable pipeline slurries. How finely these solids are ground will affect rheology on the "n and k" level, but simulation provides actual velocity profiles, power requirements, and pressure levels for scale-up. Off-design plugging and "start-up conditions" are important in pipeline operations. The same simulations also produce "snapshots" of viscous stress fields that tend to erode these structures; thus, for the first time, a tool that allows risk evaluation is available. Detailed understanding of the flow enables better characterization of wax and hydrate formation and deposition.

I am indebted to many organizations and individuals that have shaped my technical skills, professional interests, and personal perspectives over the years. In particular, I express my gratitude to the Massachusetts Institute of Technology, for providing a solid foundation in physics and mathematical analysis, and the Boeing Commercial Airplane Company, for the opportunity to learn modern grid generation and finite difference methods.

I also wish to acknowledge Halliburton Energy Services, particularly Steve Almond, Ron Morgan, and Harry Smith, and Brown & Root Energy Services, notably Raj Amin, Gee Fung, Tin Win, and Jeff Zhang, for their support in developing advanced rheological models, and especially for an environment that encourages fundamental studies and physical understanding. I am also grateful to the United States Department of Energy for its support in computer visualization, convergence acceleration, and duct geometry mapping research, provided under Grant No. DE-FG03-99ER82895. The encouragement offered by Timothy Calk, John Wilson, and James Wright, Gulf Publishing, is also kindly noted. Finally, I am indebted to my friend and colleague Mark Proett, for lending a critical and sympathetic ear to all my "crazy ideas" over the years.

Wilson C. Chin, Ph.D., M.I.T.

Houston, Texas
Email: wilsonchin@aol.com

1

Introduction:
Basic Principles and Applications

Students of fluid mechanics learn many laws of nature. For example, the Hagen-Poiseuille pipe flow formula "$Q = \pi R^4 \Delta p /(8 \mu L)$," an exact consequence of the Navier-Stokes equations, gives the steady total volume flow rate Q for a fluid with viscosity μ, flowing under a pressure drop Δp, in a circular pipe of radius R and length L. Especially significant are its dependencies; that is, doubling the pressure drop doubles the flow rate, doubling viscosity halves the flow rate, and so on. Similar Navier-Stokes solutions are obtained for other engineering applications, which also yield considerable physical insight.

However, the widely studied Navier-Stokes equations apply only to "simple" fluids like air and water, known as "Newtonian" fluids. Fortunately, a large number of practical Newtonian applications deal with important problems, for instance, external flows past airplanes, internal flows within jet engines, and free surface flows about ships, submarines, and offshore platforms. But for wide classes of fluids, unfortunately, the rules of thumb available to Newtonian flows break down, and useful design laws and operational guidelines are lost.

For example, in the context of pipe flow, the notion of a "viscosity μ" is no longer simple, even when pressure and temperature are fixed: not only does it depend on flow rate, container size and shape, but it also varies throughout the cross-section of the duct. To complicate matters, there are different classes of non-Newtonian fluids, or "rheologies," e.g., power law, Bingham plastic, Herschel-Bulkley, and literally dozens of "constitutive laws" or stress-strain relationships characterizing different types of emulsions and slurries.

Real flows can be unforgiving. For example, the fluid "seen" by a pipeline during its lifetime changes as produced oil and water fractions and composition change. Even if the rheological model remains the same, simple "flow rate versus pressure drop" statements are still not possible; for instance, when the "n and k" for a power law fluid changes, the corresponding "Q versus Δp"

1

relationship changes. Because typical rate relationships are very typically nonlinear, it is difficult to speculate, for instance, on what pump pressures might be required to initiate a given start-up flow rate in a stopped pipeline. In most cases, doubling the pressure drop will not double the total flow rate.

Non-Newtonian flows are challenging from an analysis viewpoint. Few exact solutions are available, and then, only for simple fluid models and circular pipe cross-sections. But it is not difficult to imagine subsea pipelines blocked by accumulated wax or by hydrate plugs, as shown in Figures 1-1a,b, bundled pipes with debris settlement, as illustrated in Figure 1-1c, or heavily clogged eccentric drillhole annuli, as depicted in Figure 1-1d, requiring analysis for planning or remedial work. For such geometries, there are no solutions.

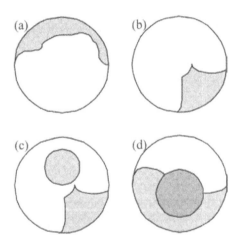

Figure 1-1. Typical clogged pipe and annular configurations.

The severity of many operational problems is worsened by inaccessibility: clogged underwater pipelines and stuck horizontal drillpipe are virtually unreachable from the surface, and remedial efforts must be performed from afar. Economic consequences, e.g., lost production in the case of pipelines, rig rental fees when not "making hole," are usually costly. These considerations drive the need for rheological planning early on. For example, "What pump pressures are required in 'worst case' flow start-up?" "What flow properties are associated with a given drilling mud, cement, or emulsion?" "How much production is lost for variously shaped plugs and obstructions?" "Can flow blockage be inferred from changes in "Q versus Δp" data?" "What kinds of annular designs are optimal for heated bundled pipelines, and how are coupled velocity and temperature fields calculated for such configurations?" "Can advanced simulation algorithms be encoded in real-time control software?"

Until recently, these questions were academic because flow models for "real world" problems, typified by those in Figure 1-1, were impossible. Not only were the partial differential equations nonlinear, making their solution unwieldy, but the geometrical boundaries where "no-slip" velocity conditions must be applied were not amenable to simple description. Engineers and designers applied questionable analysis methods, e.g., "mean hydraulic radius," "slot flow," and so on, to practical problems, often incurring serious error.

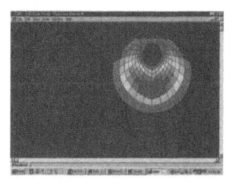

Figure 1-2a. Non-Newtonian velocity, highly eccentric annulus. (CDROM contains full-color versions of all "properties figures.")

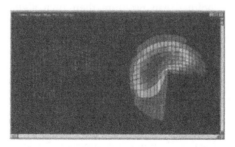

Figure 1-2b. Non-Newtonian velocity, pipe with large blockage.

My research in *Borehole Flow Modeling* was first to combine the iterative relaxation methods used to solve partial differential equations on rectangular meshes, with new approaches to curvilinear gridding and surface definition pioneered in modern topology (Chin, 1991). As a result, practical solutions for non-Newtonian flow could be computed for the first time in highly eccentric annuli, within a matter of seconds. In the present book, these original models are extended to handle pipe flows with large blockages, "multiply-connected" annular flows, in which large pipes contain multiple smaller pipes, flows with time-dependent debris deposition, and then, coupled heat transfer.

My objectives here are multi-faceted. A book devoted to numerical analysis and mathematics would lose sight of the physical problems that motivated their importance; it would also appeal to a limited audience. Thus, I emphasize the engineering and operational aspects of the motivating issues as well, focusing on *why* new, uniquely different approaches may be necessary to replace older, more "comfortable" methodologies. And whenever possible, I present applications of the methods, and offer empirical "proof" and validation, based on quantitative experimental data or qualitative observation.

But the presentations are no less rigorous: the reader interested in mathematics and computational technique will find our expositions sufficiently complete, with references to detailed work amply provided. However, to those who have seen one too many "$\tau = k (d\gamma/dt)^n$" formulas without real application, computed velocity results like those in Figures 1-2a,b, together with "typical numbers" and "snapshots" of complementary apparent viscosity, stress, and shear rate fields, will be welcome. And unlike books focusing on theory and literature surveys, our methods are immediately available for practical use, necessarily, by means of software, given the sophisticated background mathematics involved. This is all the more important because simple rules for non-Newtonian flows on complicated geometries cannot be simply given: they must be computed for each individual case and studied on a comparative basis.

While highly specialized details related to partial differential equation methods have been omitted, enough discussion is provided so that readers who are mathematically inclined can program their own algorithms based on the material presented. And because "finite difference" methods are used to define mappings and transformations, and as well, to solve all host flow equations for different rheological models, a thorough introduction (including Fortran source code) is offered that demonstrates how iterative solutions are constructed.

I also highlight the importance of analytical solutions by offering them whenever possible, emphasizing their power and elegance; but the fact that completely different methods are required, even for minor changes to duct geometry, at the same time draws attention to their inherent weaknesses. In this sense, our computational techniques are superior, because a single algorithm applies, for example, to duct cross-sections that are circular, rectangular, triangular, or perhaps, shaped as shown in Figures 1-2a,b.

Borehole Flow Modeling importantly used computational methods to show how, in horizontal and highly deviated wells, mean viscous stress obtained at the low side of highly eccentric annuli directly correlated with cuttings transport efficiency. This is so because cuttings beds are characterized by well-defined mechanical yield stresses. In subsea pipelines, stress, velocity, and other quantities may be important to removing wax deposits, loose debris, and hydrate plugs; our models provide the means to interpret flow loop research data, to develop solids deposition models that couple with our duct analysis methods, and to provide predictive means to study transient plugging and remediation.

WHY STUDY RHEOLOGY?

Many books on rheology develop constitutive equations and kinematical relationships, offering illustrations, typically showing flat "small n" velocity profiles and even flatter "plug flows" characteristic of Bingham plastics. In pipeline and annular petroleum applications, these are not very useful since the primary concerns are operational ones that focus on plugging. This book instead describes computer methods for "real world" geometries, and applications to problems like cuttings transport, sand cleaning, wax deposition, and hydrate buildup. In addition, numerous "color snapshots" of practical flowfields are given, not just for velocity, but for apparent viscosity, shear rate, viscous stress, and dissipation function distributions. In several examples, actual "numbers" are given to provide the "physical feel" that engineers need to enhance their "personal" experience with such flows. Also, general results for a number of difficult eccentric annular configurations are given for Newtonian flows.

The author believes that a book describing modern theory and numerical methods has limited value unless software is made available to industry. In the same vein, software is not very useful unless problem set-up is straightforward and fast, computations are extremely rapid, and three-dimensional color displays of all spatially varying quantities are immediately available on solution. This is imperative because engineers use such models not as a means to study rheology in itself, but as a means to understand problems that plague operational efficiency. For example, "Does viscous stress correlate with cuttings bed removal?" "As beds become less 'cohesive,' to what extent can velocity-based correlations be used?" "How can existing civil engineering 'rules of thumb' be extrapolated to oilfield pipeline dimensions?" "Can the extensions be legitimately made when the rheologies are non-Newtonian?"

I emphasize that while the algorithms are predictive and very efficient, they do have their limitations. In this book, I study steady, laminar flows only, and do not consider turbulence. Also, chemistry and thermodynamics are not considered because they are not the focus of this effort. I emphasize the role of fluid mechanics as one providing correlation parameters for debris transport and cleaning, and highlight the methods used in developing deposition and erosion models. However, the development of models specific to individual applications is a research endeavor in itself, so that specific models, consequently, are not given. Instead, qualitative results obtained from a number of client applications are offered to provide the reader with broad exposure to the potential afforded by computational rheology methods.

Nonetheless, broad usage is possible, even within these constraints. In the literature, simplifying models have been used to simulate a variety of industry problems, and it is enlightening to offer at least a partial list of applications:

- Process design for manufacturing, e.g., heat and melt flow behavior in dies, extruder screws, molds, and so on,

- Industrial manufacturing, e.g., wire coating extrusion, coatings for glass rovings, extrusions, mixing, coating, and injection molding for food and polymer processing,

- Roller coating of foil and aluminum sheets,

- Modeling power requirements and viscous drag for peanut butter and mayonnaise flows,

- Movement of "mechanically extracted meat" in food processing (after meat is removed from carcasses and bone is separated, it is compressed with "crushers," ultimately becoming "goo-like" with nonlinear characteristics).

Within the petroleum industry, rheology is important in all aspects of exploration and production, and also in oilfield development and pipeline transport. In concluding, we can list numerous applications, among them,

- In drilling vertical wells, cuttings are efficiently removed by increasing velocity, viscosity, or both, while in horizontal and highly deviated wells, cleaning efficiency correlates with bottom viscous stress in eccentric annuli,

- Flowline debris are transported as "solids in fluid" systems, or alternatively, ground slurries, with rheological considerations entering economic decisions,

- Efficient cementing and completions require a good understanding of rheology as it relates to mud displacement and pumping requirements,

- Hydraulic fracturing and stimulation involve proppant transport, with rheology dictating how well a fluid convects solid particulates and how resistant it is to pumping (one less pumping truck in the field can add significantly to profit margins!),

- In deep subsea applications, "flow assurance," i.e., the application of prevention and remediation techniques, and operating strategies, to possible flowline blockage, is used to transport crude economically,

- Accurate modeling of wax deposition and hydrate formation, and their potential for plugging flowlines, requires coupled solutions with non-Newtonian flow solvers in order to model interacting solids and fluids flowfields,

- The thermal performance of subsea bundle flowlines, involving the solution of coupled velocity and temperature fields, requires non-Newtonian flow analysis in complicated domains, and

- In offshore operations, severe slugging in risers and tiebacks is a major concern in the design and operation of deepwater production systems.

REVIEW OF ANALYTICAL RESULTS

We have satisfactorily answered "Why study rheology?" In petroleum engineering, we emphasize that "rheology" necessarily implies "computational rheology," since operational questions bearing important economic implications cannot be answered without dealing with measured constitutive relationships and actual clogged pipeline and annular borehole geometries. Before delving into our subject matter, it is useful to review several closed form solutions. These are useful because they provide important validation points for calculated results, and instructive because they show how limiting analytical methods are. For our purposes, I will not list one-dimensional, planar solutions, which have limited petroleum industry applications, but focus on pipe and annular flows in this section. Rectangular ducts will be treated later in Chapter 7.

Newtonian pipe flow. What can be simpler than flow in a pipe? In this chapter, we will find that most "sophisticated" analytical solutions are available for pipe flows only, and then, limited to just several rheological models.

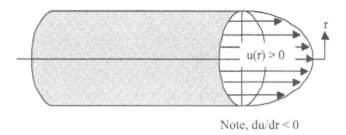

Note, du/dr < 0

Figure 1-3. Axisymmetric pipe flow.

Figure 1-3 illustrates straight, axisymmetric, pipe flow, where the axial velocity $u(r) > 0$ depends on the radial coordinate $r > 0$. With these conventions, the "shear rate" $du/dr < 0$ is negative, that is, $u(r)$ decreases as r increases. Very often, the notation $d\gamma/dt = - du/dr > 0$ is used. If the viscous shear stress τ and the shear rate are linearly related by

$$\tau = - \mu \, du/dr > 0 \qquad (1\text{-}1a)$$

where "μ" is the viscosity, a constant or temperature dependent quantity, then two simple relationships can be derived for pipe flow.

Let $\Delta p > 0$ be the (positive) pressure drop over a pipe of length L, and R be the inner radius of the pipe. Then, the radial velocity distribution satisfies

$$u(r) = [\Delta p/(4\mu L)] (R^2 - r^2) > 0 \qquad (1\text{-}1b)$$

Note that u is constrained by a "no-slip" velocity condition at $r = R$. If the product of "$u(r)$" and the infinitesimal ring area "$2\pi r\, dr$" is integrated over $(0,R)$, we obtain the volume flow rate expressed by

$$Q = \pi R^4 \Delta p/(8\mu L) > 0 \qquad (1\text{-}1c)$$

Equation (1-1c) is the well-known Hagen-Poiseuille formula for flow in a pipe. These solutions do not include unsteadiness or compressibility. These results are exact relationships derived from the Navier-Stokes equations, which govern viscous flows when the stress-strain relationships take the linear form in Equation 1-1a. We emphasize that the Navier-Stokes equations apply to Newtonian flows only, and not to more general rheological models.

Note that viscous stress (and the wall value τ_w) can be calculated from Equation 1-1a, but the following formulas can also be used,

$$\tau(r) = r\, \Delta p/2L > 0 \qquad (1\text{-}2a)$$

$$\tau_w = R\, \Delta p/2L > 0 \qquad (1\text{-}2b)$$

Equations 1-2a,b apply generally to steady laminar flows in circular pipes, and importantly, whether the rheology is Newtonian or not. But they do not apply to ducts with other cross-sections, or to annular flows, even concentric ones, whatever the fluid.

Bingham plastic. Bingham plastics satisfy a slightly modified constitutive relationship, usually written in the form,

$$\tau = \tau_y - \mu\, du/dr \qquad (1\text{-}3a)$$

where τ_y represents the yield stress of the fluid. In other words, fluid motion will not initiate until stresses exceed yield; in a moving fluid, a "plug flow" moving as a solid body is always found below a "plug radius" defined by

$$R_p = 2\tau_y L/\Delta p \qquad (1\text{-}3b)$$

The "if-then" nature of this model renders it nonlinear, despite the (misleading) linear appearance in Equation 1-3a. Fortunately, simple solutions are known,

$$u(r) = (1/\mu)\,[\{\Delta p/(4L)\}\,(R^2 - r^2) - \tau_y\,(R - r)],\ R_p \le r \le R \qquad (1\text{-}3c)$$

$$u(r) = (1/\mu)\,[\{\Delta p/(4L)\}\,(R^2 - R_p^2) - \tau_y\,(R - R_p)],\ 0 \le r \le R_p \qquad (1\text{-}3d)$$

$$Q/(\pi R^3) = [\tau_w/(4\mu)]\,[1 - 4/3\,(\tau_y/\tau_w) + 1/3\,(\tau_y/\tau_w)^4] \qquad (1\text{-}3e)$$

Power law fluids. These fluids, without yield stress, satisfy the power law model in Equation 1-4a, and the rate solutions in Equations 1-4b,c.

$$\tau = k(-du/dr)^n \qquad (1\text{-}4a)$$

$$u(r) = (\Delta p/2kL)^{1/n}\,[n/(n+1)]\,(R^{(n+1)/n} - r^{(n+1)/n}) \qquad (1\text{-}4b)$$

$$Q/(\pi R^3) = [R\Delta p/(2kL)]^{1/n}\,n/(3n+1) \qquad (1\text{-}4c)$$

Nonlinear "Q vs. Δp" graphical plots are given in Chapter 8. We emphasize that linear behavior applies to Newtonian flows exclusively.

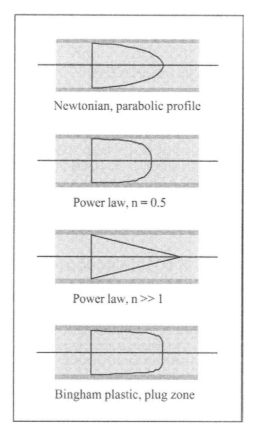

Figure 1-4. Typical non-Newtonian velocity profiles.

Herschel-Bulkley fluids. This model combines power law with yield stress characteristics, with the result that,

$$\tau = \tau_y + k(-du/dr)^n \tag{1-5a}$$

$$u(r) = k^{-1/n}(\Delta p/2L)^{-1}\{n/(n+1)\} \tag{1-5b}$$
$$\times [(R\Delta p/2L - \tau_y)^{(n+1)/n} - (r\Delta p/2L - \tau_y)^{(n+1)/n}], \ R_p \leq r \leq R$$

$$u(r) = k^{-1/n}(\Delta p/2L)^{-1}\{n/(n+1)\} \tag{1-5c}$$
$$\times [(R\Delta p/2L - \tau_y)^{(n+1)/n} - (R_p\,\Delta p/2L - \tau_y)^{(n+1)/n}], \ 0 \leq r \leq R_p$$

$$Q/(\pi R^3) = k^{-1/n}(R\Delta p/2L)^{-3}(R\Delta p/2L - \tau_y)^{(n+1)/n} \tag{1-5d}$$
$$\times [(R\Delta p/2L - \tau_y)^2 n/(3n+1) + 2\tau_y(R\Delta p/2L - \tau_y)n/(2n+1) + \tau_y^2 n/(n+1)]$$

where the plug radius R_p is again defined by Equation 1-3b.

Ellis fluids.   Ellis fluids satisfy a more complicated constitutive relationship, with the following known results,

$$\tau = - du/dr/(A + B\,\tau^{\alpha-1}) \tag{1-6a}$$

$$u(r) = A\,\Delta p\,(R^2 - r^2)/(4L) + B(\Delta p/2L)^{\alpha}\,(R^{\alpha+1} - r^{\alpha+1})/(\alpha + 1) \tag{1-6b}$$

$$Q/(\pi R^3) = A\tau_w/4 + B\,\tau_w^{\alpha}/(\alpha+3) \tag{1-6c}$$

$$= A(R\Delta p/2L)/4 + B\,(R\Delta p/2L)^{\alpha}/(\alpha+3)$$

Dozens of additional rheological models appear in the literature, but specially relevant ones will be described later in this book. Typical qualitative features of the associated velocity profiles are shown in Figure 1-4.

Annular flow solutions.   The only known exact, closed form, analytical solution is a classic one describing Newtonian flow in a concentric annulus. Let R be the outer radius, and κR be the inner radius, so that $0 < \kappa < 1$. Then, it can be shown that,

$$u(r) = [R^2\Delta p/(4\mu L)]$$
$$\times\,[\,1 - (r/R)^2 + (1 - \kappa^2)\,\log_e (r/R) / \log_e (1/\kappa)\,] \tag{1-7a}$$

$$Q = [\pi R^4\Delta p/(8\mu L)]\,[\,1 - \kappa^4 - (1 - \kappa^2)^2 / \log_e (1/\kappa)\,] \tag{1-7b}$$

For non-Newtonian flows, even for concentric geometries, numerical procedures are required, e.g., see Fredrickson and Bird (1958), Bird, Stewart and Lightfoot (1960), or Skelland (1967). The limited number of exact solutions unfortunately summarizes the state-of-the-art, and for this reason, recourse must be made to computational rheology for the great majority of practical problems.

Particulate settling.   The terminal velocity of particles is important to deposition studies and particulate settling. Because a "well-known" formula is indiscriminately used, it is instructive to point out key assumptions (and hence, restrictions) behind the result.   Newton's law "**F** = m**a**" requires that the acceleration d^2z/dt^2 propelling a mass M must equal the external force **F**.   In this case, it consists of the weight "-Mg," where g is the acceleration due to gravity, the buoyancy force "4/3 $\pi a^3 \rho_f g$," where a sphere of radius "a" and a fluid of density ρ_f are assumed, and finally, a hydrodynamic viscous drag.

Usually, a Newtonian flow is assumed for the latter, and then, in the low Reynolds number limit, for which the laminar drag becomes "$6\pi\mu aU$."   This formula also assumes a smooth sphere, and dynamic effects such as "fluttering" are ignored.   Even so, the mathematics involved in its derivation is complicated; for non-Newtonian flows, analogous closed form solutions are not available. Terminal velocity is obtained by setting $d^2z/dt^2 = 0$ and solving for "U." Needless to say, the result does not apply to non-Newtonian flows, nor to particles other than spherical, but given the lack of solutions, the result is used often anyway, although sometimes with empirical corrections. Kapfer (1973) provides examples of this common usage.

OVERVIEW OF ANNULAR FLOW

Although pipe flows precede annular ones in simplicity, and "ought" to be discussed in that order, I consider annular flows first because the mapping methods used in this book were originally developed and empirically validated for such applications. Chapters 2-5 describe three recent annular borehole flow models, first presented in *Borehole Flow Modeling*, and now reevaluated with more physical insight. They were designed to handle the special problems that arise from drilling and producing deviated and horizontal wells, e.g., cuttings transport, stuck pipe, cementing and coiled tubing. The models deal with (i) eccentric, nonrotating flow, (ii) concentric rotating flow, and (iii) recirculating heterogeneous flow. In this section, we introduce the subject of borehole annular flow, briefly discuss the capabilities of the models, and describe operational problems that benefit from detailed flow analysis.

The first model allows arbitrary eccentricity, assuming that the pipe (or casing) does not rotate. It solves the *complete* nonlinear viscous flow equations on "boundary conforming" grids, and does *not* invoke the "narrow annulus," "parallel plate," "slot flow," or "hydraulic radius" assumptions commonly used. Holes with washouts, cuttings beds, and square drill collars, for example, are easily simulated. The model is developed for Newtonian, power law, Bingham plastic, and Herschel-Bulkley fluids.

The second model permits general pipe rotation, but it is restricted to concentric geometries. For Newtonian fluids, the results are shown to be *exact* solutions to the Navier-Stokes equations; both axial and azimuthal velocities satisfy no-slip conditions for all diameter ratios. For power law flows, a narrow annulus assumption is invoked that allows us to derive explicit closed form analytical solutions for all physical quantities. These formulas are easily programmed on pocket calculators. The results are checked with our Newtonian formulas and are consistent with these exact results in the "n=1" limit.

These models assume constant density, unidirectional axial flow where applied pressure gradients are exactly balanced by viscous stresses. These flows form the majority of observed fluid motions. But in deviated wells, especially where drilling mud circulation has been temporarily interrupted, gravity segregation often causes weighting materials such as barite, fine cuttings, and cement additives to fall out of suspension. The resulting density variations and inertial effects are primarily responsible for the strange "recirculating vortex flows" that have been experimentally observed from time to time. These isolated tornado-like clusters, completely fluid-dynamical in origin, are dangerous because they impede the mainstream flow; also, they entrain drilled cuttings, and form stationary obstacles in the annulus.

Recirculating flows are stable packets of angular momentum that are wholly self-contained in a stationary envelope that sits in the midst of an axial flow. The latter flow is, effectively, blocked. Within the envelope are rotating fluid masses, some of which are roped off by closed streamlines; these highly three-dimensional flows are known to capture and trap solid particles and cuttings. The third model describes these fascinating fluid motions and identifies the controlling nondimensional parameter. Computer simulations showing their generation and growth are given, and ways to avoid or eliminate their occurrence are suggested. *These recirculating vortices have also been observed in pipelines in actual process plant case histories, and are therefore relevant to flow assurance studies in deep subsea applications.*

Although the mathematical models and numerical simulators had been available since 1987, original book publication was withheld until 1991, pending application to field and laboratory examples. Often, the required data and empirical results were either unavailable or proprietary, contributing to delays in the evaluation of the work. Experimental validation was crucial in establishing the credibility and accuracy of the computer modeling, especially because analytical solutions simply do not exist for the purposes of verification. Since numerical differencing methods, iteration and programming techniques invariably introduce additional assumptions that may be unphysical, consistency checks with empirical results were essential. These extraneous effects include truncation error, mesh dependence, and numerical viscosity.

Eventually, cuttings transport data, stuck pipe and other complementary information became available, and the desired comparisons were undertaken after some initial delay. The first applications results were published in a series of articles carried by *Offshore Magazine* beginning in 1990. An expanded "field-oriented" treatment dealing with rigsite applications is offered here in Chapter 5. This book explains in detail the mathematical models and numerical algorithms used, provides calculated examples of "difficult" annular flows, and applies the computer models to problems related in hole cleaning, stuck pipe, and cementing in deviated and horizontal wells. The extension to coupled velocity and thermal flowfields, important to analyzing "bundled pipelines" in deep subsea applications, is carried out in Chapter 6.

It is not essential to understand the details of the mathematics in order to appreciate the nuances of annular flow as uncovered by our calculations. In fact, the reader is encouraged to browse through the computed "snapshots" prior to any detailed study. Mathematics aside, the practical implications suggested by our examples will be understandable to most petroleum engineers. The experienced researcher will have little trouble programming the flow models derived here. However, practitioners may obtain algorithms from Gulf Publishing Company in the form of PC-executable software. Professionals interested in source code extensions and more sophisticated versions of the available program are encouraged to contact the author directly. All numerical algorithms are written in standard Fortran, and are readily ported to different hardware environments. Further software details are offered in Chapter 10.

REVIEW OF PRIOR ANNULAR MODELS

Annular flow analysis is important to drilling and production in deviated and horizontal wells. Different applications will be introduced and covered in detail in Chapter 5. Despite their significance, few rigorous simulation models are available for research or field use. There are several reasons for this dearth of analysis. First, the governing equations are nonlinear; this means that any useful solutions are necessarily numerical. Second, most practical annular geometries are complicated, making no-slip velocity boundary conditions difficult to enforce with any accuracy. Third, few computational algorithms are presently available for general rheologies that are stable, fast and robust.

Consequently, researchers have chosen to study simpler although less realistic models whose mathematics are at least amenable to solution. These limitations have now been overcome, to some extent through technology transfer from related disciplines. Much of our work on eccentric flow, for example, represents an extension of aerospace industry research in simulating annular-like motions in jet engine ducts. Nonlinear equations are solved, for example, using fast iterative techniques developed by aerodynamicists for shear flows. And the work of Chapter 4 on heterogeneous flows draws, in part, upon the literature of dynamic meteorology and oceanography.

The model development summarized in this book is self-contained. The complete equations of motion for a fluid having a general stress tensor (Bird, Stewart and Lightfoot, 1960; Schlichting, 1968; Slattery, 1981; Streeter, 1961) are assumed. They are solved using physical boundary conditions relevant to petroleum applications (Gray and Darley, 1980; Moore, 1974; Whittaker, 1985; Quigley and Sifferman, 1990; Govier and Aziz, 1977). The resulting formulations are solved using special relaxation methods and analytical techniques. An introduction to these methods may be found in Lapidus and Pinder (1982), Crochet *et al.* (1984), and Thompson *et al.* (1985). Let us review the existing literature on annular flow, emphasizing that no attempt is made to offer an exhaustive or comprehensive survey.

Modeling efforts may be classified into several increasingly sophisticated categories. For example, the exact solution of Fredrickson and Bird (1958) for power law fluids is among the simplest; this numerical solution applies to concentric, nonrotating power law fluids. Eccentric annular flows are more complicated, and have been modeled under limiting assumptions. To simplify the mathematics, authors assume that the annulus is "almost concentric." This "parallel plate," "narrow annulus," or "slot flow" assumption is almost universal in the petroleum literature. But the results of these investigations are appealing because they provide a convenient analytical representation of the solution using elliptic integrals. However, their usefulness is severely limited because few eccentric annuli in deviated wells are nearly concentric.

Recently, Haciislamoglu and Langlinais (1990) importantly removed this slot flow restriction by reformulating the governing equations in "bipolar coordinates." Just as circular polar coordinates imply simplifications to single well radial flow simulation, bipolar coordinates allow exact annular flow modeling of *circular* drillpipes and boreholes with arbitrary standoffs. The authors used an iterative finite difference method to model Bingham fluids, but they did not provide information on computing times and numerical stability or code portability. However, limitations on the mapping used means that the methodology *cannot* be extended to handle boreholes with cuttings beds and washouts, or noncircular drillpipes and casings with stabilizers or centralizers.

In a second important paper, Haciislamoglu and Langlinais (1990) correctly pointed out that slot flow approaches simulate radial shear only and neglect that component in the circumferential direction. These models, in other words, incorrectly use the equation for narrow concentric flow without accounting for the additional circumferential shear due to eccentricity. Another category of annular models reverts to simpler concentric flows, but allows constant speed rotation. Some of these are listed in the references. However, the solution techniques are cumbersome and not amenable for even research use. In Chapter 3, a closed form analytical solution is derived whose results agree with Savins and Wallick (1966) and Luo and Peden (1989a, 1989b).

THE NEW COMPUTATIONAL MODELS

This book focuses on the annular models conceived in *Borehole Flow Modeling*, and extended since then to include coupled velocity and thermal fields, with significant speed improvements. Also, general methodologies for non-Newtonian duct flows with arbitrary geometric cross-sections appear in print for the first time. New algorithms are presented that show how duct simulation methods can be combined with solids deposition and erosion models to describe wax buildup and hydrate formation in deep subsea flowlines.

Eccentric, nonrotating annular flow. The need for fast, stable, and accurate flow solvers for general eccentric annuli is central to drilling and production engineering. Because of mathematical difficulties, i.e., the nonlinearity of the governing equations and the complexity of most geometries, the problem is usually simplified by using unrealistic slot flow assumptions. Even then, the unwieldy elliptic integrals that result shed little physical insight into what remains of the problem. Moreover, the integrals require intensive computations, further decreasing their usefulness in field applications.

In Chapter 2, annular cross-sections containing eccentric circles are permitted. But importantly, the borehole contour may be modified "point by point" to simulate the effects of cuttings beds or wall deformations due to erosion and swelling. The pipe (or casing) contour may be likewise modified, for example, to model square drill collars or stabilizers and centralizers. Narrow annulus and slot flow assumptions are *not* invoked. The analysis model handles Newtonian, power law, Bingham plastic, and Herschel-Bulkley fluids. In all cases, the formulation satisfies "no-slip" velocity boundary conditions exactly at *all* solid surfaces. Since the appearance of *Borehole Flow Modeling*, underbalanced drilling has gained in popularity. Changes in algorithm design now permit the modeling of "velocity slip," crucial modeling foam-based muds.

Our model is derived from first principles using the exact equations of continuum mechanics. These equations are written in coordinates natural to the annular geometry considered. Then second-order accurate solutions for the axial velocity field, shear rate, shear stress, apparent viscosity, Stokes product, and dissipation function are obtained. These solutions make use of recent developments in boundary conforming grid generation, e.g., Thompson *et al.* (1985) and relaxation methods (Lapidus and Pinder, 1982; Crochet *et al.*, 1984).

Special graphics software allows computed results to be overlaid on the annular geometry itself, so that physical trends can be visually correlated with position. The unconditionally stable iteration process requires ten seconds on typical Pentium class personal computers. Efficient coding allows the executable code to reside in less than 100K RAM. One organization, in fact, has incorporated both mapping and flow solvers successfully in real-time control software. Note that the Fortran code is portable and compatible with all machine environments. The computer program, now Windows-based, is written for generalists and uses "plain English" menus requiring only the experience of novice petroleum engineers. Graphics capabilities have been significantly upgraded, and are described in Chapter 6.

The only restriction is our assumption of a stationary nonrotating pipe. This is not overly severe in field applications, since the intended application of the model was anticipated in horizontal turbodrilled wells. Also, rotational effects will not be important when the tangential pipe speed is small compared to the average axial speed; for such problems, estimates are easily obtained using the dynamic model of Chapter 3.

Eccentric annuli with thermal effects. In deep subsea production, heated pipelines are sometimes carried within larger pipes containing crude. Heat is necessary to prevent wax deposition and hydrate formation. Unlike the velocity simulations above, where fluid attributes like "n" and "k" are constant, these properties now depend on temperature, which itself satisfies a thermal boundary value problem dictated by the heating line and ocean environment. Chapter 6 extends our methodology to such flows, and illustrative calculations are performed, showing the differences that arise for two "bundled" geometries.

Concentric, rotating annular flow. Rotational effects are important when drilling at low flow rates, or when rotating the casing during cementing. For eccentric annuli, all three velocity components will be nonlinearly coupled. Although numerical simulation is possible, the simultaneous solution of three velocity equations requires computing resources not usually available at user locations. For this reason, the eccentric model of Chapter 2 is restricted to zero rotation. Constant speed rotation, by contrast, can be treated quite generally for concentric annuli. However, the usual solution techniques do not lend themselves to simple use; the final equations are *implicit* and require iteration.

In Chapter 3, the restriction to power law fluids in narrow annuli is made. Then a simple but powerful application of the Mean Value Theorem of differential calculus allows us to derive closed form solutions for the relatively complicated problem. Note that the resulting problem still contains *four* coupled no-slip conditions, two for each of the axial and circumferential velocities. The solutions are shown to be consistent with an exact solution of the Navier-Stokes equations for Newtonian flow, which does not bear any limiting geometric restrictions. Formulas for apparent viscosity, stress, deformation and dissipation function versus "r" are given. The solutions are *explicit* in that they require no iteration. A Fortran graphics algorithm is also described that conveniently plots and tabulates desired solutions, without requiring additional investment in graphical software and hardware.

Recirculating annular and pipe flows. The recirculating flows described earlier are interesting and fascinating in their own right. But they may be responsible for operational problems. In drilling, the presence of dynamically stable fluid-dynamic obstacles in the annular mainstream means that cuttings transport will be impaired. These stationary structures may affect bed buildup both upstream and downstream. In cementing applications, their presence in the mud or the cement slurry would suggest ineffective mud displacement. This implies poor zonal isolation and the need for corrective squeeze cementing.

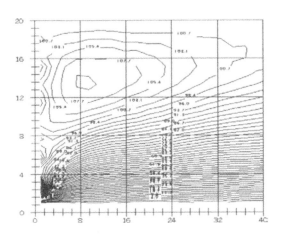

Figure 1-5. Recirculating vortex zone.

These vortex structures have recently been identified in pipelines in process plant applications. In one published study, the recirculation zone entrained abrasive particles that eroded to pipe over time. Originally, engineers speculated that high streamwise velocities were to blame; however, subsequent analysis clearly showed that recirculation was the culprit. In any event, the transverse extent of any recirculation bubbles likely to be present is the ultimate solution sought. Thus, the model presented in Chapter 4 solves for streamline shapes and boundaries. Velocities, stresses, and pressures can in principle be obtained from streamline patterns. The fractional blockage inferred from the presence of any closed streamlines provides a qualitative danger indicator for cuttings removal and cement displacement.

Duct flows with arbitrary geometries. The second half of this book is devoted to non-Newtonian pipe flow: not "simple" flows, e.g., Equations 1-1 to 1-6, but flows with significant clogging. Our focus is practical, dwelling on "Q vs. Δp" relationships for pipe clogs typical of wax deposition and hydrate formation. Such results are especially relevant to "start-up" requirements, when pipelines have stopped production temporarily and fluids have gelled. As with annular flow, we again calculate detailed spatial flowfields without approximation, except to the extent that our flow equations are discretized and solved using finite difference methods. Exact no-slip conditions are again used, and quantities like apparent viscosity, shear rate, viscous stress, and dissipation function, are obtained by post-processing the calculated velocity field. The new duct simulator is described in detail in Chapter 7, and applications to wax deposition, hydrate growth, and erosion by the convected flow, are developed in Chapter 8. Additional advanced topics are covered in Chapters 9 and 10.

PRACTICAL APPLICATIONS

We introduce some practical applications for the annular flow simulators developed in Chapters 2, 3, and 4, and also the pipe flow simulator described in Chapter 7. These field applications are listed and briefly discussed.

Cuttings transport in deviated wells. The most important operational problem confronting drillers of deviated and horizontal wells is cuttings transport and bed formation. Many excellent experimental studies have been performed by industry and university groups, but the results are often confusing. For example, take eccentric inclined annular flows. Velocity, which plays an important role in vertical holes, has little value as a correlation parameter beyond 30° deviation. But mean viscous stress turns out to be the parameter of significance; the right threshold value will erode cuttings beds formed at the bottom of highly eccentric annuli. Extensive field data and computational results support this view in Chapter 5.

What role do rotational viscometer measurements taken at the surface play in downhole applications? *None, because downhole shear rates, which change from case to case, are not known a priori.* Or consider pipe rotation. It turns out that concentric Newtonian flows, often used as the basis for convenient experiments, have no bearing on real world problems. In this singular limit, both the axial and azimuthal velocities decouple dynamically, and experimental observations cannot be extrapolated to other situations.

Spotting fluids and stuck pipe. A related problem is the severe one dealing with stuck pipe. Typically, spotting fluids are used in combination with mechanical jarring motions to free immobile pipe. These impulsive transient flows can also be approximately analyzed since the acceleration and pressure gradient terms in the momentum equation have like physical dimensions. What parameter governs spotting fluid effectiveness? How is that quantity related to lubricity? These questions are also addressed in Chapter 5.

Coiled tubing return flow. Sands and fines are often produced in flowing wells. They are removed by injecting non-Newtonian foams delivered by coiled tubing forced into the well. The debris is then transported up the return annulus inside the production tubing, a process not unlike the movement of drilled cuttings. Here, the weight and small diameter of the metal tubing (typically, 1 to 2 inches, O.D.) render the annular geometry highly eccentric. This problem is also suited to the model of Chapter 2.

Cementing. Proper primary cementing creates the fluid seals needed to produce formation fluids properly. Improper procedures often lead to expensive, difficult squeeze cementing jobs. Mud that remains in the hole often does so because the cement velocity profile is hydrodynamically unstable, admitting viscous fingering and laminar flow breakdown. How are stable velocity profiles selected? How are dangerous "recirculation zones" that impede effective mud displacement eliminated? How does casing rotation alter the state of stress in the mud? These questions are addressed in Chapters 2-5.

Improved well planning. Well planning involves mud pump selection and drilling fluid properties calculations. In vertical wells with concentric annuli, these are straightforward. But simple questions require complicated answers for eccentric annular spaces, and particularly when they contain non-Newtonian fluids. "Can the pump operate through a range of mud weights and flow rates for a long horizontal well?" The dependence of flow rate on pressure gradient, of course, is nonlinear; and just as problematic is the apparent viscosity distribution, which depends on pressure gradient as well as annular geometry.

Borehole stability. Borehole stability depends on several factors, principally mud chemistry and elastic states of stress. But annular flow can be important. For example, rapid velocities or surface stresses can erode borehole walls and promote washouts in unconsolidated sands. Drilling muds may also prove abrasive since they carry drilled cuttings that impinge into the formation.

Wellbore heat generation. Temperature effects can be important in drilling. While heat generation due to internal friction is small, overall temperature increases in a closed system may be significant for large circulation times. These may affect the thermal stability and thinning of oil-base muds. Heat generation may be important to temperature log interpretation. To calculate formation temperature correctly from measurements obtained while drilling, it is necessary to correct for the component of temperature due to total circulation time and the rheological effects of the drilling fluid used.

Modern pipe flow analysis. I have alluded to new non-Newtonian flow simulation methods, and their potential application to wax deposition, hydrate formation, and possible erosion by the fluid stream.

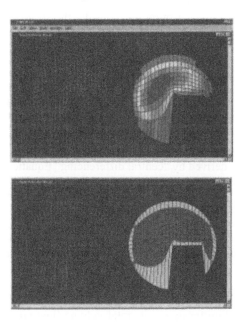

Figure 1-6. Velocity fields past hydrate plug.

If a single picture is worth a thousand words, then Figure 1-6, for power law (top) and Bingham plastic (bottom) fluids, must be worth more. Here, an approximate 25% area decrease in flow area significantly lowers the volume throughput that would obtain in an unclogged pipe. I challenge the reader, at this point, to speculate on the amount of the decrease! Again, this book significantly expands upon previously published articles in *Offshore Magazine* (Chin, 1990a,b,c; Chin, 1991), and *Formation Invasion* (Chin, 1995), dealing with borehole fluids that have invaded the formation. New developments include significant increases in convergence rate, upgraded graphical capabilities, general duct flow analysis, and multiply-connected zones.

PHILOSOPHY OF NUMERICAL MODELING

Reservoir engineers and structural dynamicists, for example, routinely use advanced finite difference and finite element methods. But drillers have traditionally relied upon simpler handbook formulas and tables that are convenient at the rigsite. Simulation methods are powerful, to be sure, but they have their limitations. This section explains the pitfalls and the philosophy one must adopt in order to bring state-of-the-art techniques to the field. Importantly, we emphasize that *numerical methods do not always yield exact answers.* But more often than not, they produce excellent *trend information* that is useful in practical application. For our purposes, consider the steady, concentric annular flow of a Newtonian fluid (Bird *et al.*, 1960). The governing equations are

$$d^2u(r)/dr^2 + r^{-1} \, du/dr = (1/\mu) \, dp/dz \qquad (1\text{-}8a)$$

$$u(R_i) = u(R_o) = 0 \qquad (1\text{-}8b)$$

In Equations 1-8a,b, u(r) is the annular speed satisfying no-slip conditions at the inner and outer radii, R_i and R_o. Here, the viscosity μ and the applied pressure gradient dp/dz are known constants. The exact solution was given earlier.

Let us examine the consequences of a numerical solution. A "second-order accurate" scheme is derived by "central differencing" Equation 1-8a as follows,

$$(u_{j-1} - 2u_j + u_{j+1})/(\Delta r)^2 + (u_{j+1} - u_{j-1})/2r_j\Delta r = (1/\mu) \, dp/dz \quad (1\text{-}9a)$$

where u_j refers to u(r) at the jth radial node at the r_j location, j being an ordering index. Equation 1-9a can be evaluated at any number of interior nodes for the mesh length Δr. The resulting difference equations, when augmented by

$$u_1 = u_{jmax} = 0 \qquad (1\text{-}9b,c)$$

using Equation 1-8b, form a tridiagonal system of j_{max} unknowns that lends itself to simple solution for u_j and its total volume flow rate. For our first run, we assumed $R_i = 4$ inch, $R_o = 5$ inch, dp/dz = - 0.0005 psi/in and $\mu = 2$ cp. Computed flow rates as functions of mesh density are given in Table 1-1.

# Meshes	GPM	% Error
2	783	25
3	929	11
4	980	6
5	1003	4
10	1035	1
20	1042	0
30	1044	0
100	1045	0

Table 1-1. Volume flow rate versus mesh number.

Note how the "100 mesh" solution is almost exact; but the "10 mesh" solution for flow rate, which is ten times faster to compute, is satisfactory for engineering purposes. Now let us double the viscosity μ and recompute the solution. The gpm's so obtained decrease exactly by a factor of two, and the dependence on viscosity is certainly brought out very clearly. However, the *trend information* relating changes in gpm to those in μ are accurately captured even for coarse meshes. So, sometimes fine meshes are unnecessary. Similar comments apply to the pressure gradient dp/dz.

It is clear that the exact value of u(r) is mesh dependent; the finer the mesh, the better the answer. In some applications, it may be essential to find, through trial and error, a mesh distribution that leads to the exact solution or that is consistent with real data in some engineering sense. From that point on, "what if" analyses may be performed accurately with confidence. This rationale is used in reservoir engineering, where history matching with production data plays a crucial role in estimating reserves. For other applications, the exact numbers may not be as important as qualitative trends of different physical parameters. For example, how does hole eccentricity affect volume flow rate for a prescribed pressure gradient? For a given annular geometry, how does a decrease in the power law exponent affect velocity profile curvature?

In structural engineering, it is well known that *uncalibrated* finite element analyses can accurately pinpoint *where* cracks are likely to form even though the computed stresses may not be correct. For such qualitative objectives, the results of a numerical analysis may be accepted "as is" provided the calculated numbers are not literally interpreted. Agreement with exact solutions, of course, is important; but often it is the very lack of such analytical solutions itself that motivates numerical alternatives. Thus, while consistency with exact solutions is desirable, in practice it is through the use of *comparative solutions* that computational methods offer their greatest value.

For annular flows and pipe flows in ducts having general cross-sectional geometries, this philosophy is appropriate because there are no analytical solutions or detailed laboratory measurements with which to establish standards for comparison. One should be satisfied as long as the solutions agree roughly with field data; the real objective, remember, aims at establishing *trends* with respect to *changes* in parameters like fluid rheology, flow rate, and hole eccentricity. We will show through extensive computations and correlation with empirical data that the models developed in Chapters 2, 3, and 4 are correct and useful in this engineering sense. The ultimate acid test lies in field applications, and these are addressed in Chapter 5.

I emphasize that the eccentric flow of Chapter 2, the original thrust of this research, is by no means as simple as the above example might suggest. In Equation 1-8a, the unknown speed u(r) depends on a single variable "r" only. In Chapter 2, the velocity depends on two cross-sectional coordinates x and y; this leads to a partial differential equation that is also generally nonlinear.

The "two-point" boundary conditions in Equation 1-8b are therefore replaced by no-slip velocity conditions enforced along two general arbitrary curves representing the borehole and pipe contours. To implement these no-slip conditions accurately, "boundary conforming meshes" must be used that provide high resolution in tight spaces. To be numerically efficient, these meshes must be *variable* with respect to all coordinate directions. The difference equations solved on such host meshes must be solved *iteratively*; for unlike Equations 1-9a,b,c, which apply to Newtonian flows with constant viscosities, the power law, Bingham plastic, and Herschel-Bulkley fluids considered in this book satisfy nonlinear equations with problem-dependent apparent viscosities. The algorithms must be *fast, stable, and robust*; they must produce solutions without straining computing resources. Finally, computed solutions must be physically correct; this is the final arbiter that challenges all numerical simulations.

REFERENCES

Bird, R.B., Stewart, W.E., and Lightfoot, E.N., *Transport Phenomena*, John Wiley and Sons, New York, 1960.

Chin, W.C., "Advances in Annular Borehole Flow Modeling," *Offshore Magazine*, February 1990, pp. 31-37.

Chin, W.C., "Exact Cuttings Transport Correlations Developed for High Angle Wells," *Offshore Magazine*, May 1990, pp. 67-70.

Chin, W.C., "Annular Flow Model Explains Conoco's Borehole Cleaning Success," *Offshore Magazine*, October 1990, pp. 41-42.

Chin, W.C., "Model Offers Insight into Spotting Fluid Performance," *Offshore Magazine*, February 1991, pp. 32-33.

Chin, W.C., *Borehole Flow Modeling in Horizontal, Deviated, and Vertical Wells*, Gulf Publishing Company, Houston, Texas, 1992.

Chin, W.C., "Eccentric Annular Flow Modeling for Highly Deviated Boreholes," *Offshore Magazine*, Aug. 1993.

Chin, W.C., *Formation Invasion, with Applications to Measurement-While-Drilling, Time Lapse Analysis, and Formation Damage*, Gulf Publishing Company, Houston, Texas, 1995.

Crochet, M.J., Davies, A.R., and Walters, K., *Numerical Simulation of Non-Newtonian Flow*, Elsevier Science Publishers B.V., Amsterdam, 1984.

Davis, C.V., and Sorensen, K.E., *Handbook of Applied Hydraulics*, McGraw-Hill, New York, 1969.

Fredrickson, A.G., and Bird, R.B., "Non-Newtonian Flow in Annuli," *Ind. Eng. Chem.*, Vol. 50, 1958, p. 347.

Govier, G.W., and Aziz, K., *The Flow of Complex Mixtures in Pipes*, Robert Krieger Publishing, New York, 1977.

Gray, G.R., and Darley, H.C.H., *Composition and Properties of Oil Well Drilling Fluids*, Gulf Publishing Company, Houston, 1980.

Haciislamoglu, M., and Langlinais, J., "Non-Newtonian Fluid Flow in Eccentric Annuli," 1990 ASME Energy Resources Conference and Exhibition, New Orleans, January 14-18, 1990.

Haciislamoglu, M., and Langlinais, J., "Discussion of Flow of a Power-Law Fluid in an Eccentric Annulus," *SPE Drilling Engineering*, March 1990, p. 95.

Iyoho, A.W., and Azar, J.J., "An Accurate Slot-Flow Model for Non-Newtonian Fluid Flow Through Eccentric Annuli," *Society of Petroleum Engineers Journal*, October 1981, pp. 565-572.

King, R.C., and Crocker, S., *Piping Handbook*, McGraw-Hill, New York, 1973.

Langlinais, J.P., Bourgoyne, A.T., and Holden, W.R., "Frictional Pressure Losses for the Flow of Drilling Mud and Mud/Gas Mixtures," SPE Paper No. 11993, 58th Annual Technical Conference and Exhibition of the Society of Petroleum Engineers, San Francisco, October 5-8, 1983.

Lapidus, L., and Pinder, G., *Numerical Solution of Partial Differential Equations in Science and Engineering*, John Wiley and Sons, New York, 1982.

Luo, Y., and Peden, J.M., "Flow of Drilling Fluids Through Eccentric Annuli," Paper No. 16692, SPE Annual Technical Conference and Exhibition, Dallas, September 27-3, 1987.

Luo, Y., and Peden, J.M., "Laminar Annular Helical Flow of Power Law Fluids, Part I: Various Profiles and Axial Flow Rates," SPE Paper No. 020304, December 1989.

Luo, Y., and Peden, J.M., "Reduction of Annular Friction Pressure Drop Caused by Drillpipe Rotation," SPE Paper No. 020305, December 1989.

Moore, P. L., *Drilling Practices Manual*, PennWell Books, Tulsa, 1974.

Perry, R.H., and Chilton, C.H., *Chemical Engineer's Handbook*, McGraw-Hill, New York, 1973.

Quigley, M.S., and Sifferman, T.R., "Unit Provides Dynamic Evaluation of Drilling Fluid Properties," *World Oil*, January 1990, pp. 43-48.

Savins, J.G., "Generalized Newtonian (Pseudoplastic) Flow in Stationary Pipes and Annuli," *Petroleum Transactions*, AIME, Vol. 213, 1958, pp. 325-332.

Savins, J.G., and Wallick, G.C., "Viscosity Profiles, Discharge Rates, Pressures, and Torques for a Rheologically Complex Fluid in a Helical Flow," *A.I.Ch.E. Journal*, Vol. 12, No. 2, March 1966, pp. 357-363.

Schlichting, H., *Boundary Layer Theory*, McGraw-Hill, New York, 1968.

Skelland, A.H.P., *Non-Newtonian Flow and Heat Transfer*, John Wiley & Sons, New York, 1967.

Slattery, J.C., *Momentum, Energy, and Mass Transfer in Continua*, Robert E. Krieger Publishing Company, New York, 1981.

Streeter, V.L., *Handbook of Fluid Mechanics*, McGraw-Hill, New York, 1961.

Thompson, J.F., Warsi, Z.U.A., and Mastin, C.W., *Numerical Grid Generation*, Elsevier Science Publishing, New York, 1985.

Uner, D., Ozgen, C., and Tosun, I., "Flow of a Power Law Fluid in an Eccentric Annulus," *SPE Drilling Engineering*, September 1989, pp. 269-272.

Vaughn, R.D., "Axial Laminar Flow of Non-Newtonian Fluids in Narrow Eccentric Annuli," *Society of Petroleum Engineers Journal*, December 1965, pp. 277-280.

Whittaker, A., *Theory and Application of Drilling Fluid Hydraulics*, IHRDC Press, Boston, 1985.

Yih, C.S., *Fluid Mechanics*, McGraw-Hill, New York, 1969.

Zamora, M., and Lord, D.L., "Practical Analysis of Drilling Mud Flow in Pipes and Annuli," SPE Paper No. 4976, 49th Annual Technical Conference and Exhibition of the Society of Petroleum Engineers, Houston, October 6-9, 1974.

2

Eccentric, Nonrotating, Annular Flow

Numerical solutions for the nonlinear, two-dimensional axial velocity field, and its corresponding stress and shear rate distributions, are obtained for eccentric annular flow in an inclined borehole. The homogeneous fluid is assumed to be flowing unidirectionally in a wellbore containing a nonrotating drillstring. The unconditionally stable algorithm used draws upon finite difference relaxation methods (Lapidus and Pinder, 1982; and Crochet, Davies and Walters, 1984), and contemporary methods in differential geometry and boundary conforming grid generation (Thompson, Warsi and Mastin, 1985).

Slot flow, narrow annulus, and parallel plate assumptions are not invoked. The cross-section may contain conventional concentric or nonconcentric circular drillpipes and boreholes. But importantly, the hole and pipe contours may be arbitrarily modified "point by point" to simulate the effects of square drill collars, centralizers, stabilizers, thick cuttings beds, washouts, and general side wall deformations due to swelling and erosion. "Equivalent hydraulic radii" approximations are never used.

The overall formulation, which applies to general rheologies, is specialized to Newtonian flows, power law fluids, Bingham plastics, and Herschel-Bulkley flows. In all instances, no-slip velocity boundary conditions are satisfied exactly at all solid surfaces. Detailed spatial solutions and cross-sectional plots for local annular velocity, apparent viscosity, two components each of viscous stress and shear rate, Stokes product and heat generation due to fluid friction, are presented for a large number of annular geometries. Net volume flow rates are also given.

Calculated results are displayed using a special "character-based" text graphics program that overlays computed quantities on the annular cross-section itself, thus facilitating the physical interpretation and visual correlation of numerical quantities with annular position. For most annular geometries of practical interest, mesh generation requires approximately five seconds of computing time on Pentium machines. Once the host mesh is available, any number of "what if" scenarios for differing rheologies or net flow rates can be efficiently evaluated, these simulations again requiring five seconds. This chapter derives the basic ideas from first principles and explains them mathematically. However, the reader who is more interested in practical applications may, without loss of continuity, proceed directly to those sections.

25

THEORY AND MATHEMATICAL FORMULATION

The equations governing general fluid motions in three spatial dimensions are available from many excellent textbooks (Bird, Stewart and Lightfoot, 1960; Streeter, 1961; Schlichting, 1968; and, Slattery, 1981). We will cite these equations without proof. Let u, v and w denote Eulerian fluid velocities, and F_z, F_y and F_x denote body forces, in the z, y and x directions, respectively, where (z,y,x) are Cartesian coordinates. Also, let ρ be the constant fluid density and p be the pressure; and denote by S_{zz}, S_{yy}, S_{xx}, S_{zy}, S_{yz}, S_{xz}, S_{zx}, S_{yx} and S_{xy} the nine elements of the general extra stress tensor $\underline{S}$. If t is time and ∂'s represent partial derivatives, the complete equations of motion obtained from Newton's law and mass conservation are,

Momentum equation in z:

$$\rho \, (\partial u/\partial t + u \, \partial u/\partial z + v \, \partial u/\partial y + w \, \partial u/\partial x) =$$
$$= F_z - \partial p/\partial z + \partial S_{zz}/\partial z + \partial S_{zy}/\partial y + \partial S_{zx}/\partial x \qquad (2\text{-}1)$$

Momentum equation in y:

$$\rho \, (\partial v/\partial t + u \, \partial v/\partial z + v \, \partial v/\partial y + w \, \partial v/\partial x) =$$
$$= F_y - \partial p/\partial y + \partial S_{yz}/\partial z + \partial S_{yy}/\partial y + \partial S_{yx}/\partial x \qquad (2\text{-}2)$$

Momentum equation in x:

$$\rho \, (\partial w/\partial t + u \, \partial w/\partial z + v \, \partial w/\partial y + w \, \partial w/\partial x) =$$
$$= F_x - \partial p/\partial x + \partial S_{xz}/\partial z + \partial S_{xy}/\partial y + \partial S_{xx}/\partial x \qquad (2\text{-}3)$$

Mass continuity equation:

$$\partial u/\partial z + \partial v/\partial y + \partial w/\partial x = 0 \qquad (2\text{-}4)$$

Rheological flow models. These equations apply to all Newtonian and non-Newtonian fluids. In continuum mechanics, the most common class of empirical models for incompressible, isotropic fluids assumes that $\underline{S}$ can be related to the rate of deformation tensor $\underline{D}$ by a relationship of the form

$$\underline{S} = 2 \, N(\Gamma) \, \underline{D} \qquad (2\text{-}5)$$

where the elements of $\underline{D}$ are

$$D_{zz} = \partial u/\partial z \qquad (2\text{-}6)$$
$$D_{yy} = \partial v/\partial y \qquad (2\text{-}7)$$
$$D_{xx} = \partial w/\partial x \qquad (2\text{-}8)$$
$$D_{zy} = D_{yz} = (\partial u/\partial y + \partial v/\partial z)/2 \qquad (2\text{-}9)$$

$$D_{zx} = D_{xz} = (\partial u/\partial x + \partial w/\partial z)/2 \qquad (2\text{-}10)$$

$$D_{yx} = D_{xy} = (\partial v/\partial x + \partial w/\partial y)/2 \qquad (2\text{-}11)$$

In Equation 2-5, $N(\Gamma)$ is the "apparent viscosity" satisfying

$$N(\Gamma) > 0 \qquad (2\text{-}12)$$

$\Gamma(z,y,x)$ being a scalar functional of u, v and w defined by the tensor operation

$$\Gamma = \{\, 2 \text{ trace } (\underline{\mathbf{D}} \bullet \underline{\mathbf{D}}) \,\}^{1/2} \qquad (2\text{-}13)$$

Unlike the constant laminar viscosity in classical Newtonian flow, the apparent viscosity depends on the details of the particular problem being considered, e.g., the rheological model used, the exact annular geometry occupied by the fluid, the applied pressure gradient or the net volume flow rate. Also, it varies with the position (z,y,x) in the annular domain. Thus, single measurements obtained from viscometers may not be meaningful in practice.

Power law fluids. These considerations are general. To fix ideas, we examine one important and practical simplification. The Ostwald-de Waele model for two-parameter "power law" fluids assumes that

$$N(\Gamma) = k\, \Gamma^{n-1} \qquad (2\text{-}14a)$$

where the "consistency factor" k and the "fluid exponent" n are constants. Such power law fluids are "pseudoplastic" when $0 < n < 1$, Newtonian when $n = 1$, and "dilatant" when $n > 1$. Most drilling fluids are pseudoplastic. In the limit $(n=1, k=\mu)$, Equation 2-14a reduces to the Newtonian model with $N(\Gamma) = \mu$, where μ is the constant laminar viscosity; in this classical limit, stress is directly proportional to the rate of strain. Only for Newtonian flows is total volume flow rate a linear function of applied pressure gradient.

Yield stresses. Power law and Newtonian fluids respond instantaneously to applied pressure and stress. But if the fluid behaves as a rigid solid until the net applied stresses have exceeded some known critical yield value, say S_{yield}, then Equation 2-14a can be generalized by writing

$$N(\Gamma) = k\, \Gamma^{n-1} + S_{yield}/\Gamma \text{ if } \{1/2 \text{ trace } (\underline{\mathbf{S}} \bullet \underline{\mathbf{S}})\}^{1/2} > S_{yield}$$

$$\underline{\mathbf{D}} = 0 \text{ if } \{1/2 \text{ trace } (\underline{\mathbf{S}} \bullet \underline{\mathbf{S}})\}^{1/2} < S_{yield} \qquad (2\text{-}14b)$$

In this form, Equation 2-14b rigorously describes the general Herschel-Bulkley fluid. When the limit $(n=1, k=\mu)$ is taken, the first equation becomes

$$N(\Gamma) = \mu + S_{yield}/\Gamma \text{ if } \{1/2 \text{ trace } (\underline{\mathbf{S}} \bullet \underline{\mathbf{S}})\}^{1/2} > S_{yield} \qquad (2\text{-}14c)$$

This is the Bingham plastic model, where μ is now the "plastic viscosity." Annular flows containing fluids with nonzero yield stresses are more difficult to analyze, both mathematically and numerically, than those marked by zero yield.

This is so because there may co-exist "dead" (or "plug") and "shear" flow regimes with internal boundaries that must be determined as part of the solution. Even though "n = 1," Bingham fluid flows are essentially nonlinear.

For now, we restrict our discussion to Newtonian and power law flows, that is, to fluids without yield stresses. For flows whose velocities do not depend on the axial coordinate z, and which further satisfy v = w = 0, the functional Γ in Equation 2-14a takes the form

$$\Gamma = [\ (\partial u/\partial y)^2 + (\partial u/\partial x)^2\ ]^{1/2} \qquad (2\text{-}15)$$

so that Equation 2-14a becomes

$$N(\Gamma) = k\ [\ (\partial u/\partial y)^2 + (\partial u/\partial x)^2\ ]^{(n-1)/2} \qquad (2\text{-}16)$$

The apparent viscosity reduces to the conventional "$N(\Gamma) = k\ (\partial u/\partial y)^{(n-1)}$" formula for one-dimensional, parallel plate flows considered in the literature.

When both independent variables y and x for the cross-section are present, as in the case for eccentric annular flow, significant mathematical difficulty arises. For one, the *ordinary* differential equation for annular velocity in simple concentric geometries becomes a *partial* differential equation (PDE). And whereas the former requires boundary conditions at two points, the PDE requires no-slip boundary conditions along two arbitrarily closed curves. The nonlinearity of the governing PDE and the irregular annular geometry only compound these difficulties.

Borehole configuration. The configuration considered is shown in Figure 2-1. A drillpipe (or casing) and borehole combination is inclined at an angle α relative to the ground, with $\alpha = 0°$ for horizontal and $\alpha = 90°$ for vertical wells. Here "z" denotes any point within the annular fluid; Section "AA" is a cut taken normal to the local z axis. Figure 2-2 resolves the vertical body force due to gravity at "z" into components parallel and perpendicular to the axis. Figure 2-3 provides a detailed picture of the annular cross-section at Section "AA."

Now specialize the above equations to downhole flows. In Figures 2-1, 2, and 3, we have aligned z, which increases downward, with the axis of the borehole. The axis may be inclined, varying from $\alpha = 0°$ for horizontal to $90°$ for vertical holes. The plane of the variables (y,x) is perpendicular to the z-axis, and (z,y,x) are mutually orthogonal Cartesian coordinates. The body forces due to the gravitational acceleration g can be resolved into components

$$F_z = \rho\ g \sin \alpha \qquad (2\text{-}17)$$

$$F_x = -\rho\ g \cos \alpha \qquad (2\text{-}18)$$

$$F_y = 0 \qquad (2\text{-}19)$$

If we now assume that the drillpipe does not rotate, the resulting flow can only move in a direction parallel to the borehole axis. This requires that the velocities v and w vanish. Therefore,

$$v = w = 0 \qquad\qquad (2\text{-}20)$$

Since the analysis applies to constant density flows, we obtain

$$\partial\rho/\partial t = 0 \qquad\qquad (2\text{-}21)$$

Equations 2-4, 2-20, and 2-21 together imply that the axial velocity $u(y,x,t)$ does not depend on z. And, if we further confine ourselves to steady laminar flow, that is, to flows driven by axial pressure gradients that do not vary in time, we find that

$$u = u(y,x) \qquad\qquad (2\text{-}22)$$

depends at most on two independent variables, namely the cross-sectional coordinates y and x.

In the case of a concentric drillpipe and borehole, it is more convenient to collapse y and x into a radial coordinate r. This is accomplished by using the definition $r = (x^2 + y^2)^{1/2}$. For general eccentric flows, the lack of similar algebraic transformations drives the use of grid generation methods. Next, substitution of Equations 2-20 and 2-22 into Equations 2-1, 2 and 3 leads to

$$0 = \rho\, g \sin \alpha \, - \partial p/\partial z \, + \partial S_{zy}/\partial y \, + \partial S_{zx}/\partial x \qquad\qquad (2\text{-}23)$$

$$0 = - \,\partial p/\partial y \qquad\qquad (2\text{-}24)$$

$$0 = - \,\rho g \cos \alpha \, - \partial p/\partial x \qquad\qquad (2\text{-}25)$$

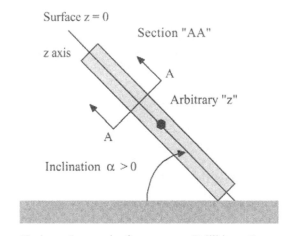

Figure 2-1. Borehole configuration.

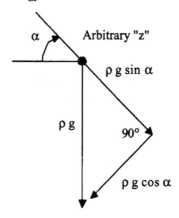

Figure 2-2. Gravity vector components.

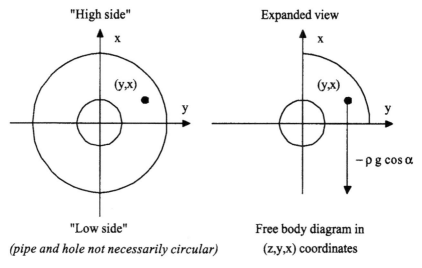

Figure 2-3. Gravity vector components.

If we introduce, without loss of generality, the pressure separation of variables

$$P = P(z,x) = p - z\rho g \sin \alpha + x\rho g \cos \alpha \qquad (2\text{-}26)$$

we can replace Equations 2-23, 2-24 and 2-25 by the single equation

$$\partial S_{zy}/\partial y + \partial S_{zx}/\partial x = \partial P/\partial z = \text{constant} \qquad (2\text{-}27)$$

where the constant pressure gradient $\partial P/\partial z$ is prescribed.

Recall the definitions of the deformation tensor elements given in Equations 2-6 to 2-11 and the fact that $\underline{S} = 2N\underline{D}$ to rewrite Equation 2-27 as

$$\partial (N \, \partial u/\partial y)/ \, \partial y + \partial (N \, \partial u/\partial x)/ \, \partial x = \partial P/\partial z \qquad (2\text{-}28)$$

Here $N(\Gamma)$ is, without approximation, given by the nonlinear equation

$$N(\Gamma) = k \left[(\partial u/\partial y)^2 + (\partial u/\partial x)^2 \right]^{(n-1)/2} \qquad (2\text{-}29)$$

Equations 2-28 and 2-29 comprise the entire system to be solved along with general no-slip velocity boundary conditions at drillpipe and borehole surfaces.

It is important, for the purposes of numerical analysis, to recognize how Equation 2-28 can be written as a nonlinear Poisson equation, that is,

$$\partial^2 u/\partial y^2 + \partial^2 u/\partial x^2 = [\partial P/\partial z + (1\text{-}n)N(\Gamma)(u_y{}^2 u_{yy}$$
$$+ 2u_y u_x u_{yx} + u_x{}^2 u_{xx})/(u_y{}^2 + u_x{}^2)] / N(\Gamma) \qquad (2\text{-}30)$$

In this form, conventional solution techniques for elliptic equations can be employed. These include iterative techniques as well as direct inversion methods. The nonlinear terms in the square brackets, for example, can be evaluated using latest values in a successive approximations scheme.

Also various algebraic simplifications are possible. For some values of n, particularly those near unity, these nonlinear terms may represent negligible higher order effects if the "1-n" terms are small in a dimensionless sense compared with pressure gradient effects. For small n, the second derivative terms on the right side may be unimportant since such flows contain flat velocity profiles. Crochet, Davies and Walters (1984), which deals exclusively with non-Newtonian flows, presents discussions on different limit processes.

For the above limits, the principal effects of nonlinearity can be modeled using the simpler and stabler Poisson model that results, one not unlike the classical equation for Newtonian flow. Of course, the apparent viscosity that acts in concert with the driving pressure gradient is still variable, nonlinear, and dependent on geometry and rate. For such cases only (and all solutions obtained in this fashion should be checked *a posteriori* against the full equation) we have

$$\partial^2 u/\partial y^2 + \partial^2 u/\partial x^2 \approx N(\Gamma)^{-1} \, \partial P/\partial z \qquad (2\text{-}31)$$

where Equation 2-29 is retained in its entirety. This approximation does not always apply. But the strong influence of local geometry on annular velocity (e.g., low bottom speeds in eccentric holes regardless of rheology or flow rate) suggests that any errors incurred by using Equation 2-31 may be insignificant. This simplification is akin to the "local linearization" used in nonlinear aerodynamics; in any case, the exact geometry of the annulus is always kept.

As noted earlier, borehole temperature may be important in drilling and production, but is probably not; thus, most studies neglect heat generation by internal friction. Internal heat generation may affect local fluid viscosity since n and k depend on temperature. One way to estimate its importance is through the strength of the temperature sources distributed within the annulus. The starting point is the energy equation for the temperature $T(z,y,x,t)$. Even if velocity is steady in time, temperature does not have to be. For example, in a closed system, temperatures will increase if the borehole walls do not conduct heat away as quickly as it is produced; weak heat production can lead to large increases in T over time. Thus, whether or not frictional heat production is significant, reduces to a matter of time scale and temperature boundary conditions. If the increases are significant, the changes of viscosity as functions of T must be modeled. This leads to mathematical complications; if the laminar viscosity $\mu = \mu(T)$ in Newtonian flow depends on temperature, say, then the momentum and energy equations will couple through this dependence.

We will not consider this coupling yet. We assume that all rheological input parameters are constants, so that our velocities obtain independently of T. Now the energy equation for T contains a positive definite quantity Φ called the "dissipation function" that is the distributed source term responsible for local heat generation. In general, it takes the form

$$\Phi(z,y,x) = S_{zz}\partial u/\partial z + S_{yy}\partial v/\partial y + S_{xx}\partial w/\partial x$$
$$+ S_{zy}(\partial u/\partial y + \partial v/\partial z) + S_{zx}(\partial u/\partial x + \partial w/\partial z)$$
$$+ S_{yx}(\partial v/\partial x + \partial w/\partial y) \tag{2-32}$$

Applying assumptions consistent with the foregoing analysis, we obtain

$$\Phi = N(\Gamma)\ \{(\partial u/\partial y)^2 + (\partial u/\partial x)^2\} > 0 \tag{2-33}$$

where, as before, we use Equation 2-29 for the apparent viscosity in its entirety. Equation 2-33 shows that velocity gradients, not magnitudes, contribute to temperature increases.

In our computations, we provide values of local viscous stresses and their corresponding shear rates. These stresses are the rectangular components

$$S_{zy} = N(\Gamma)\ \partial u/\partial y \tag{2-34}$$
$$S_{zx} = N(\Gamma)\ \partial u/\partial x \tag{2-35}$$

The shear rates corresponding to Equations 2-34 and 2-35 are $\partial u/\partial y$ and $\partial u/\partial x$ respectively. These quantities are useful for several reasons. They are physically important in estimating the efficiency with which fluids in deviated wells remove cuttings beds having specified mechanical properties. From the numerical analysis point of view, they allow checking of computed solutions for physical consistency (e.g., high values at solid surfaces, zeros within plug flows) and required symmetries. We next discuss mathematical issues regarding computational grid generation and numerical solution. These ideas are highlighted because we solve the complete boundary value problem, satisfying no-slip velocity conditions exactly, without simplifying the annular geometry.

BOUNDARY CONFORMING GRID GENERATION

In many engineering problems, a judicious choice of coordinate systems simplifies calculations and brings out the salient physical features more transparently than otherwise. For example, the use of cylindrical coordinates for single well problems in petroleum engineering leads to elegant "radial flow" results that are useful in well testing. Cartesian grids, on the other hand, are preferred in simulating oil and gas flows from rectangular fields.

The annular geometry modeling considered here is aimed at eccentric flows with cuttings beds, arbitrary borehole wall deformations, and unconventional drill collar or casing-centralizer cross-sections. Obviously, simple coordinate transforms are not readily available to handle arbitrary domains of flow. Without resorting to crude techniques, for instance, applying boundary conditions along mean circles and squares, or invoking "slot flow" assumptions, there has been no real reason for optimism until recently.

Fortunately, results from differential geometry allow us to construct "boundary conforming, natural coordinates" for computation. These general techniques extend classical ideas on conformal mapping. They have accelerated progress in simulating aerospace flows past airfoils and cascades, and are only beginning to be applied in the petroleum industry. Thompson, Warsi and Mastin (1985) provides an excellent introduction to the subject.

To those familiar with conventional analysis, it may seem that the choice of (y,x) coordinates in Equation 2-31 is "unnatural." After all, in the limit of a concentric annulus, the equation does not reduce to a radial formulation. But our use of such coordinates was motivated by the new gridding methods which, like classical conformal mapping, are founded on Cartesian coordinates. The approach in essence requires us to solve first a set of nonlinearly coupled, second-order PDEs. In particular, the equations

$$(y_r^2 + x_r^2)\, y_{ss} - 2(y_s y_r + x_s x_r)\, y_{sr} + (y_s^2 + x_s^2)\, y_{rr} = 0 \qquad (2\text{-}36)$$

$$(y_r^2 + x_r^2)\, x_{ss} - 2(y_s y_r + x_s x_r)\, x_{sr} + (y_s^2 + x_s^2)\, x_{rr} = 0 \qquad (2\text{-}37)$$

are considered with special mapping conditions related to the annular geometry. These are no simpler than the original flow equations, but they importantly introduce a first step that does not require solution on complicated domains.

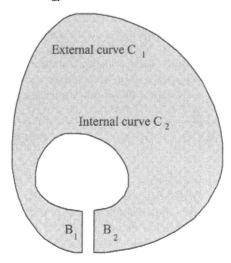

Figure 2-4a. Irregular physical (y,x) plane.

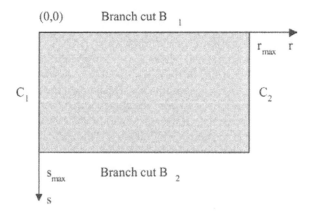

Figure 2-4b. Rectangular computational plane.

Equations 2-36 and 2-37 are importantly solved on simple rectangular (r,s) grids. Once the solution is obtained, the results for $x(r,s)$ and $y(r,s)$ are used to generate the metric transformations needed to reformulate the physical equations for u in (r,s) coordinates. The flow problem is then solved in these rectangular computational coordinates using standard numerical methods. These new coordinates implicitly contain all the details of the input geometry, providing fine resolution in tight spaces as needed. To see why, we now describe briefly the boundary conditions used in the mapping. Figures 2-4a and 2-4b indicate how a general annular region would map into a rectangular computational space under the proposed scheme.

Again the idea rests with special computational coordinates (r,s). A discrete set of "user-selected" physical coordinates (y,x) along curve C_1 in Figure 2-4a is specified along the straight line r = 0 in Figure 2-4b. Similarly, (y,x) values obtained from curve C_2 in Figure 2-4a are specified along r = r_{max} in Figure 2-4b. Values for (y,x) chosen along "branch cuts" $B_{1,2}$ in Figure 2-4a are required to be single-valued along edges s = 0 and s = s_{max} in Figure 2-4b.

With (y,x) prescribed along the rectangle of Figure 2-4b, Equations 2-36 and 2-37 for y(r,s) and x(r,s) can be numerically solved. Once the solution is obtained, the one-to-one correspondences between all physical points (y,x) and computational points (r,s) are known. The latter is the domain chosen for numerical computation for annular velocity. Finite difference representations of the no-slip conditions "u = 0" that apply along C_1 and C_2 of Figure 2-4a are very easily implemented in the rectangle of Figure 2-4b.

At the same time, the required modifications to the governing equation for u(y,x) are modest. For example, the simplified Equation 2-31 becomes

$$(y_r^2 + x_r^2)\, u_{ss} - 2(y_s y_r + x_s x_r)\, u_{sr}$$
$$+ (y_s^2 + x_s^2)\, u_{rr} = (y_s x_r - y_r x_s)^2\, \partial P/\partial z\, /N(\Gamma) \qquad (2\text{-}38)$$

whereas the result for the Equation 2-30 requires additional terms. For Equation 2-38 and its exact counterpart, the velocity terms in the apparent viscosity $N(\Gamma)$ of Equation 2-29 transform according to

$$u_y = (x_r u_s - x_s u_r)/(y_s x_r - y_r x_s) \qquad (2\text{-}39)$$
$$u_x = (y_s u_r - y_r u_s)/(y_s x_r - y_r x_s) \qquad (2\text{-}40)$$

These relationships are also used to evaluate the dissipation function.

NUMERICAL FINITE DIFFERENCE SOLUTION

We have transformed the computational problem for the annular speed u from an awkward one in the physical plane (y,x) to a simpler one in (r,s) coordinates, where the irregular domain becomes rectangular. In doing so, we introduced the intermediate problem dictated by Equations 2-36 and 2-37. When solutions for y(r,s) and x(r,s) and their corresponding metrics are available, Equation 2-38, which is slightly more complicated than the original Equation 2-31, can be solved conveniently using existing "rectangle-based" methods without compromising the annular geometry.

Equations 2-36 and 2-37 were solved by rewriting them as a single vector equation, employing simplifications from complex variables, and discretizing the end equation using second-order accurate formulas. The finite difference equations are then reordered so that the coefficient matrix is sparse, banded, and computationally efficient. Finally, the "Successive Line Over Relaxation" (SLOR) method was used to obtain the solution in an implicit and iterative manner. The SLOR scheme is unconditionally stable on a linearized von Neumann basis (for example, see Lapidus and Pinder, 1982).

Mesh generation requires approximately five seconds of computing time on Pentium machines. Once the transformations for $y(r,s)$ and $x(r,s)$ are available for a given annular geometry, Equations 2-38 to 2-40 can be solved any number of times for different applied pressure gradients, volume flow rate constraints, or fluid rheology models, without recomputing the mapping.

Because Equation 2-38 is similar to Equations 2-36 and 2-37, the same procedure was used for its solution. These iterations converged quickly and stably because the meshes used were smooth. When solutions for the velocity field $u(r,s)$ are available (these also require five seconds), simple inverse mapping relates each computed "u" with its unique image in the physical (y,x) plane. With $u(y,x)$ and its spatial derivatives known, post-processed quantities like $N(\Gamma)$, S_{zy}, S_{zx}, their corresponding shear rates, apparent viscosities and Φ are easily calculated and displayed in physical (y,x) coordinates.

Drilling and production engineers recognize that flow properties in eccentric annuli correlate to some extent with annular position (e.g., low bottom speeds regardless of rheology). Our text based graphical display software projects $u(y,x)$ and all post-processed quantities on the annular geometry. This helps visual correlation of computed physical properties or inferred characteristics (e.g., "cuttings transport efficiency" and "stuck pipe probability") with annular geometry quickly and efficiently. These highly visual outputs, plus sophisticated color graphics, together with the speed and stability of the scheme, promote an understanding of annular flow in an interactive, real-time manner.

Finally, we return to fluids with non-zero yield stresses. In general, there may exist internal boundaries separating "dead" (or "plug") and "shear" flow regimes. These unknown boundaries must be obtained as part of the solution. In free surface theory for water waves, or in shock-fitting methods for gasdynamic discontinuities, explicit equations are written for the boundary curve and solved with the full equations. These approaches are complicated. Instead, the "shock capturing" method for transonic flows with embedded discontinuities was used to capture these zones naturally during iterations. The conditions in Equations 2-14b,c were added to the "zero yield" code. This entailed tedious "point by point" testing during the computations, where the inequalities were evaluated with latest available solutions. But flows with plugs converge faster than flows without, because fewer matrix setups and inversions (steps necessary in computing shearing motions) are required once plugs develop.

DETAILED CALCULATED RESULTS

We first discuss results for four annular geometries containing power law fluids. In particular, we consider (i) a fully concentric annular cross-section to establish a baseline reference; (ii) the same concentric combination blocked by a thick cuttings bed; (iii) a highly eccentric annular flow with the circular pipe displaced within a circular hole without a cuttings bed; and (iv) a circular borehole containing a square drill collar.

Again the contours are not restricted to circles and squares, since the algorithm efficiently solves any combination of closed curves. One set of reference flow conditions will be calculated as the basis for comparison. These comparisons allow us to validate the physical consistency of our computed results. The simplicity of the input required to run the program is emphasized, as well as the highly visual format of all calculated quantities. These presently include axial velocity, apparent viscosity, two rectangular components of viscous stress and shear rate, Stokes product and dissipation function.

Example 1. Fully Concentric Annular Flow

The program requests input radii and center coordinates for circles that need not be concentric. Once entered, the (y,x) coordinates of both circles are displayed in tables. If the borehole and drillpipe contours require modification, e.g., to model cuttings beds or square drill collars, new coordinates entered at the keyboard replace those displayed. Alternatively, the annular contours may be "drawn" using an on-line text editor, whose results are read and interpolated.

We consider a 2-inch-radius pipe centered in a 5-inch-radius borehole. Assuming the first input option, the program displays the resultant geometry and provides an indication of relative dimensions, as shown in Figure 2-5a. For portability, all graphical input and output files appear as ASCII text, and do not require special display hardware or software. The program checks for input errors by asking the user to verify that the pipe is wholly contained within the hole. This visual check ensures that contours do not overlap; it is important when eccentric circles are modeled or when there is significant borehole wall deformation. If the geometry is realistic, automatic mesh generation proceeds, with all grid parameters chosen internally and transparently to the user.

The mapping requires five seconds on Pentium machines, assuming 24 "circumferential" and 10 "radial" grids. Since the mesh is variable, providing high resolution in tight spaces where large physical gradients are expected, and since the central differences used are second-order accurate, this more than suffices for most purposes. When the iterations are completed, the mesh is displayed. The mesh for our concentric annulus in Figure 2-5b is concentric.

Although our formulation for mesh generation problem was undertaken in (y,x) coordinates, it is clear that the end result must be the radial grid one anticipates. The power of the method is the extension of "radial" to complicated geometries.

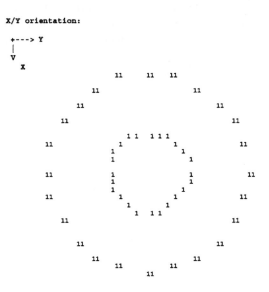

Figure 2-5a. Concentric circular pipe and hole.

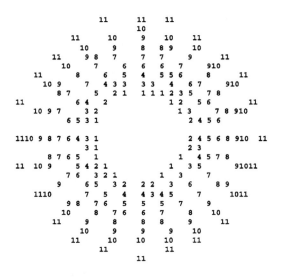

Figure 2-5b. Computed radial mesh system.

For a given input geometry, the mesh is determined only once; any number of physical simulations, to include changes to applied pressure gradient, total flow rate or fluid rheology, can be performed on that mesh. The program next asks if the working fluid is Newtonian, power law, Bingham plastic, or Herschel-Bulkley. For Newtonian flows, only the laminar viscosity is entered; the power law model can also be used with n = 1, but this results in slightly lengthier calculations. For nonlinear power law fluids, one specifies "n" and "k." For Bingham plastics and Herschel-Bulkley fluids, the yield stress is additionally required.

In this example, a power law fluid is assumed with n = 0.724 and a consistency factor of .1861E-04 lbf $\sec^n/in^2$. The axial pressure gradient is .3890E-02 psi/ft. The iterative solution to the axial velocity equation on the assumed mesh requires five seconds of computing time on Pentium machines. Results for axial velocity, apparent viscosity, two rectangular stress and shear rate components, Stokes product and dissipation function, are always displayed on the annular geometry as shown in Figures 2-5c to 2-5g. This allows convenient correlation of physical trends with annular location.

For plotting convenience, the first two significant digits of the dependent variable in the system of units used are printed (exact magnitudes are available in tabulated output). In Figure 2-5c, the first two digits of axial velocity in "in/sec" are displayed; lines of constant velocity are obtained by connecting numbers having like values. We emphasize that our solutions are not "corrected" or "mesh calibrated" by analytical methods. The computed results are displayed "as is," using internally selected mapping parameters. The objective is not so much an exact solution in the analytical sense, but accurate *comparative* solutions for a set of runs. This limited objective is more relevant to field applications, where few analytical solutions are available.

In Figure 2-5c, the 0's found at both inner and outer circular boundaries indicate that no-slip conditions have been properly and exactly satisfied. Reference to tabulated results shows that all expected symmetries are adequately reproduced by the numerical scheme. We emphasize that the ASCII character-based plotting routine provides only approximate results. "Missing numbers" and lack of symmetry are due to decimal truncation, array normalization, and character and line spacing issues. The plotter is intended as an inexpensive visualization tool that is universally portable. For more precise displays, commercial software and hardware packages are recommended.

The exact results in any event are available in output files. Typical results for the axial speed U are shown in Table 2-1. The "circumferential grid block index" is given in the left column, and corresponding coordinates appear just to the right. The index takes on a value of "1" at the "bottom middle" of any particular closed contour, and increases clockwise to "24."

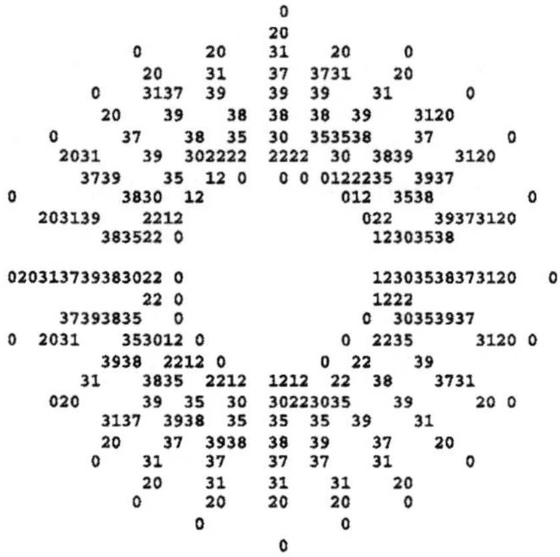

```
                              0
                             20
            0       20       31    20     0
               20    31      37  3731    20
           0  3137   39      39   39    31       0
              20     39      38   38  38  39    3120
          0   37     38   35  30  353538    37        0
        2031    39   302222   2222  30  3839    3120
        3739     35   12 0    0 0 0122235  3937
    0        3830  12               012  3538            0
    203139    2212                  022    39373120
       383522 0                     12303538

020313739383022 0                   12303538373120   0
          22 0                       1222
    37393835    0                   0 30353937
  0 2031      353012 0              0  2235       3120 0
        3938   2212 0           0  22    39
     31     3835  2212      1212   22  38    3731
    020       39   35   30   30223035     39       20 0
           3137  3938    35   35  35  39   31
            20    37  3938    38  39    37    20
          0   31     37      37  37    31          0
             20     31      31   31   20
           0        20      20   20     0
                     0       0
                            0
```

Figure 2-5c. Annular velocity.

The results in Table 2-1 for Contours No. 1 and 11 demonstrate that "no-slip" conditions are rigorously enforced at all solid boundaries. Here, the maximum speeds are found along Contour No. 6 (note the 38's and 39's in Figure 2-5c). Because the flow is concentric, all U's along this contour must be identical; this is always satisfied to the third decimal place. For internal nodes 4 to 20, the U's are identical to four places. The computed annular volume flow rate is 457.8 gal/min. Computed results for the apparent viscosity are plotted in Figure 2-5d; representative values are tabulated in Table 2-2.

Table 2-1
Example 1: Annular Velocity (in/sec)

```
Results for pipe/collar boundary, Contour No. 1:
 #  1  X= .8000E+01  Y= .6000E+01  U= .0000E+00
 #  2  X= .7932E+01  Y= .5482E+01  U= .0000E+00
 #  3  X= .7732E+01  Y= .5000E+01  U= .0000E+00
 #  4  X= .7414E+01  Y= .4586E+01  U= .0000E+00
          Results for Contour No.  6:
 #  1  X= .9149E+01  Y= .6000E+01  U=-.3873E+02
 #  2  X= .9045E+01  Y= .5184E+01  U=-.3873E+02
 #  3  X= .8732E+01  Y= .4423E+01  U=-.3874E+02
 #  4  X= .8231E+01  Y= .3769E+01  U=-.3875E+02
 #  5  X= .7578E+01  Y= .3267E+01  U=-.3875E+02
Results for borehole annular boundary, Contour No. 11:
 #  1  X= .1100E+02  Y= .6000E+01  U= .0000E+00
 #  2  X= .1083E+02  Y= .4706E+01  U= .0000E+00
 #  3  X= .1033E+02  Y= .3500E+01  U= .0000E+00
 #  4  X= .9536E+01  Y= .2464E+01  U= .0000E+00
```

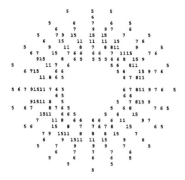

Figure 2-5d. Apparent viscosity.

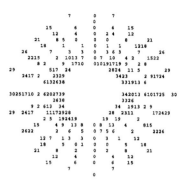

Figure 2-5f. Stress "AppVisc × dU(y,x)/dy."

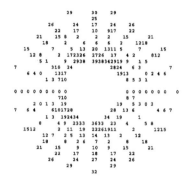

Figure 2-5e. Stress "AppVisc × dU(y,x)/dx."

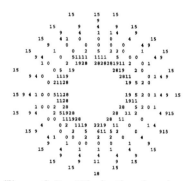

Figure 2-5g. Dissipation function.

Unlike Newtonian flows, the apparent viscosity here varies with position, changing from problem to problem. For this example, it is largest near the center of the annulus. Results for the viscous stresses "Apparent Viscosity × dU(y,x)/dx" and "Apparent Viscosity × dU(y,x)/dx" are given in Figures 2-5e and 2-5f and Tables 2-3 and 2-4.

Table 2-2
Example 1: Apparent Viscosity (lbf sec/in²)

```
Results for pipe/collar boundary, Contour No. 1:
#  7   X= .6000E+01   Y= .4000E+01   AppVisc= .5558E-05
#  8   X= .5482E+01   Y= .4068E+01   AppVisc= .5558E-05
#  9   X= .5000E+01   Y= .4268E+01   AppVisc= .5558E-05
# 10   X= .4586E+01   Y= .4586E+01   AppVisc= .5558E-05
# 11   X= .4268E+01   Y= .5000E+01   AppVisc= .5558E-05
               Results for Contour No.  7:
#  7   X= .6000E+01   Y= .2541E+01   AppVisc= .1577E-04
#  8   X= .5105E+01   Y= .2659E+01   AppVisc= .1577E-04
#  9   X= .4271E+01   Y= .3005E+01   AppVisc= .1577E-04
Results for borehole annular boundary, Contour No. 11:
# 11   X= .1670E+01   Y= .3500E+01   AppVisc= .5790E-05
# 12   X= .1170E+01   Y= .4706E+01   AppVisc= .5790E-05
# 13   X= .1000E+01   Y= .6000E+01   AppVisc= .5790E-05
# 14   X= .1170E+01   Y= .7294E+01   AppVisc= .5790E-05
```

Table 2-3
Example 1: Stress "AppVisc × dU(y,x)/dx" (psi)

```
Results for pipe/collar boundary, Contour No. 1:
#  1   X= .8000E+01   Y= .6000E+01   Stress=-.4267E-03
#  2   X= .7932E+01   Y= .5482E+01   Stress=-.3864E-03
#  3   X= .7732E+01   Y= .5000E+01   Stress=-.3460E-03
               Results for Contour No.  6:
#  4   X= .8231E+01   Y= .3769E+01   Stress=-.4616E-04
#  5   X= .7578E+01   Y= .3267E+01   Stress=-.3260E-04
#  6   X= .6817E+01   Y= .2952E+01   Stress=-.1686E-04
Results for borehole annular boundary, Contour No. 11:
# 11   X= .1670E+01   Y= .3500E+01   Stress=-.2660E-03
# 12   X= .1170E+01   Y= .4706E+01   Stress=-.2967E-03
# 13   X= .1000E+01   Y= .6000E+01   Stress=-.3072E-03
# 14   X= .1170E+01   Y= .7294E+01   Stress=-.2967E-03
```

Table 2-4
Example 1: Stress "AppVisc × dU(y,x)/dy" (psi)

```
Results for pipe/collar boundary, Contour No. 1:
#  8   X= .5482E+01   Y= .4068E+01   Stress= .3858E-03
#  9   X= .5000E+01   Y= .4268E+01   Stress= .3459E-03
# 10   X= .4586E+01   Y= .4586E+01   Stress= .2824E-03
               Results for Contour No.  6:
# 22   X= .8231E+01   Y= .8231E+01   Stress=-.4622E-04
# 23   X= .8732E+01   Y= .7577E+01   Stress=-.3270E-04
# 24   X= .9045E+01   Y= .6816E+01   Stress=-.1681E-04
Results for borehole annular boundary, Contour No. 11:
#  1   X= .1100E+02   Y= .6000E+01   Stress=-.5091E-05
#  2   X= .1083E+02   Y= .4706E+01   Stress=-.7929E-04
#  3   X= .1033E+02   Y= .3500E+01   Stress=-.1535E-03
```

Table 2-5

Example 1: Dissipation Function (lbf/(sec × sq in))

```
       Results for pipe/collar boundary, Contour No. 1:
#  6   X= .6518E+01   Y= .4068E+01   DissipFn= .2870E-01
#  7   X= .6000E+01   Y= .4000E+01   DissipFn= .2870E-01
#  8   X= .5482E+01   Y= .4068E+01   DissipFn= .2870E-01
                Results for Contour No.  7:
#  9   X= .4271E+01   Y= .3005E+01   DissipFn= .5235E-04
# 10   X= .3554E+01   Y= .3554E+01   DissipFn= .5236E-04
# 11   X= .3005E+01   Y= .4271E+01   DissipFn= .5236E-04
# 12   X= .2659E+01   Y= .5105E+01   DissipFn= .5237E-04
Results for borehole annular boundary, Contour No. 11:
# 12   X= .1170E+01   Y= .4706E+01   DissipFn= .1629E-01
# 13   X= .1000E+01   Y= .6000E+01   DissipFn= .1629E-01
# 14   X= .1170E+01   Y= .7294E+01   DissipFn= .1629E-01
# 15   X= .1670E+01   Y= .8500E+01   DissipFn= .1629E-01
```

The plotting routine omits the signs of these stresses for visual clarity; signs and exact magnitudes are available from tabulated results. This test case assuming concentric flow is important for numerical validation. The dU(y,x)/dx stress is symmetric with respect to the horizontal center line, as required, and vanishes there; similarly, the dU(y,x)/dy stress is symmetric with respect to the vertical center line and is zero there. These physically correct results appear as the result of properly converged iterations. Finally, results for the dissipation function are shown in Figure 2-5g and Table 2-5. Note how the greatest heat generation occurs at the pipe surface and at the borehole wall; there is minimal dissipation at the midpoint of the annulus.

Example 2. Concentric Pipe and Borehole in the Presence of a Cuttings Bed

---

For comparison, consider the same annular geometry used before, that is, a 2-inch-radius pipe located within a concentric 5-inch-radius borehole. But when the program requests modifications to the outer contour, we overwrite five of the bottom coordinates to simulate a flat cuttings bed. The bed height is half of the distance up the annular cross-section. This blockage should reduce the "457.8 gal/min" obtained in Example 1 for the unblocked annulus. First, the program generates the grid in Figure 2-6a, which conforms to the top of the cuttings bed. For comparison with Example 1, we again assume n = 0.7240 and a "k" of .1861E-04 lbf $\sec^n/in^2$. The pressure gradient is the same .3890E-02 psi/ft. Figure 2-6b shows that the maximum velocities at the bottom are less than one-half of those at the top. This trend is well known qualitatively, but the program allows us to obtain exact velocities everywhere without making unrealistic "slot flow" assumptions. The low velocities adjacent to the bed imply that cuttings transport will not be very efficient just above it, and that stuck pipe is possible.

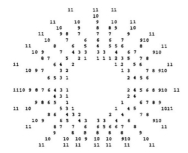

Figure 2-6a. Mesh system.

Figure 2-6b. Annular velocity.

Figure 2-6d. Stress "AppVisc × dU(y,x)/dx."

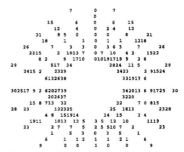

Figure 2-6e. Stress "AppVisc × dU(y,x)/dy."

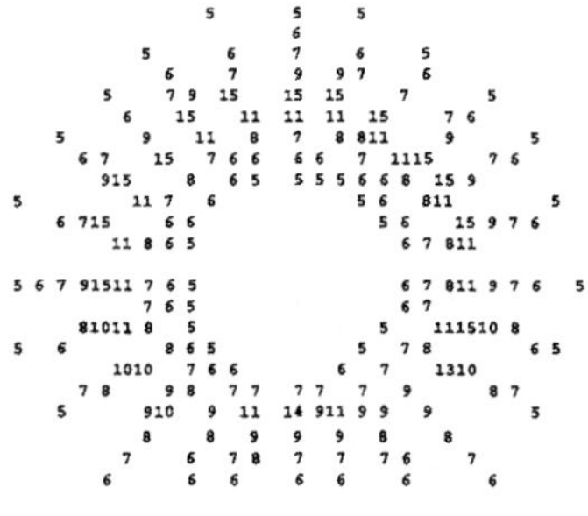

Figure 2-6f. Dissipation function.

Figure 2-6c. Apparent viscosity.

The program can also be used to determine the mud type needed to increase bottom velocities or viscous stresses to acceptable levels. The computed "as is" volume flow rate is 366.7 gal/min, much less than the 457.8 gal/min obtained for the unblocked flow of Example 1. This decreased value is consistent with physical intuition. Numerical results along the vertical line of symmetry are given in Table 2-6. These numbers represent upper and lower velocity profiles; again, note the exact implementation of no-slip conditions.

Table 2-6
Example 2: Annular Velocity (in./sec)

```
Velocity Profile (Vertical "X" coordinate increases downward)

                        Upper Annulus
  # 13  X= .1000E+01  Y= .6000E+01  U= .0000E+00
  # 13  X= .1441E+01  Y= .6000E+01  U=-.2016E+02
  # 13  X= .1843E+01  Y= .6000E+01  U=-.3189E+02
  # 13  X= .2209E+01  Y= .6000E+01  U=-.3756E+02
  # 13  X= .2541E+01  Y= .6000E+01  U=-.3917E+02
  # 13  X= .2844E+01  Y= .6000E+01  U=-.3871E+02
  # 13  X= .3120E+01  Y= .6000E+01  U=-.3590E+02
  # 13  X= .3372E+01  Y= .6000E+01  U=-.3050E+02
  # 13  X= .3601E+01  Y= .6000E+01  U=-.2259E+02
  # 13  X= .3810E+01  Y= .6000E+01  U=-.1235E+02
  # 13  X= .4000E+01  Y= .6000E+01  U= .0000E+00
                        Lower Annulus
  #  1  X= .8000E+01  Y= .6000E+01  U= .0000E+00
  #  1  X= .8096E+01  Y= .6000E+01  U=-.4868E+01
  #  1  X= .8202E+01  Y= .6000E+01  U=-.8927E+01
  #  1  X= .8319E+01  Y= .6000E+01  U=-.1207E+02
  #  1  X= .8447E+01  Y= .6000E+01  U=-.1420E+02
  #  1  X= .8588E+01  Y= .6000E+01  U=-.1528E+02
  #  1  X= .8744E+01  Y= .6000E+01  U=-.1532E+02
  #  1  X= .8914E+01  Y= .6000E+01  U=-.1414E+02
  #  1  X= .9102E+01  Y= .6000E+01  U=-.1140E+02
  #  1  X= .9309E+01  Y= .6000E+01  U=-.6785E+01
  #  1  X= .9536E+01  Y= .6000E+01  U= .0000E+00
```

Fluid viscosity is important to cuttings transport, and plays a dominant role in near-vertical holes. For example, in Newtonian flows the drag force acting on a slowly moving small particle is proportional to the product of viscosity and the relative speed. In power law flows, the apparent viscosity varies with space, but a similar correlation may apply. The apparent viscosity (among other parameters) is a useful qualitative indicator of cuttings mobility. Figure 2-6c shows the distribution of apparent viscosity for this problem, and Table 2-7 gives typical numerical values along the outer borehole/cuttings bed contour.

Table 2-7

Example 2: Apparent Viscosity (lbf sec/in.^{2})

```
Results for borehole annular boundary, Contour No. 11:
  #  1  X= .9536E+01  Y= .6000E+01  AppVisc= .6895E-05
  #  2  X= .9536E+01  Y= .5000E+01  AppVisc= .6545E-05
  #  3  X= .9536E+01  Y= .4000E+01  AppVisc= .6196E-05
  #  4  X= .9535E+01  Y= .2464E+01  AppVisc= .6421E-05

                           .

  # 19  X= .6000E+01  Y= .1100E+02  AppVisc= .5789E-05
  # 20  X= .7294E+01  Y= .1083E+02  AppVisc= .5793E-05
  # 21  X= .8500E+01  Y= .1033E+02  AppVisc= .5875E-05
  # 22  X= .9535E+01  Y= .9535E+01  AppVisc= .6421E-05
  # 23  X= .9536E+01  Y= .8000E+01  AppVisc= .6196E-05
  # 24  X= .9536E+01  Y= .7000E+01  AppVisc= .6545E-05
```

Axial velocity and apparent viscosity play direct roles in the dynamics of individual cuttings in near-vertical holes. The stability of the beds formed by particles that have descended to the lower side of the annulus in horizontal or highly deviated holes is also of interest. In this respect, the viscous fluid shear stresses acting at the surface of the cuttings bed are important. If they exceed the bed yield stress, then it is likely that the bed will erode. The program can be used to determine which powers and consistency factors are needed to erode beds with known mechanical yield properties. There are two relevant components of viscous fluid stress, namely, "Apparent Viscosity $\times$ dU(y,x)/dx" and "Apparent Viscosity $\times$ dU(y,x)/dy."

Table 2-8

Example 2: Stress "AppVisc $\times$ dU(y,x)/dx" (psi)

```
Results for borehole annular boundary, Contour No. 11:
  #  1  X= .9536E+01  Y= .6000E+01  Stress= .2068E-03
  #  2  X= .9536E+01  Y= .5000E+01  Stress= .2508E-03
  #  3  X= .9536E+01  Y= .4000E+01  Stress= .2888E-03
  #  4  X= .9535E+01  Y= .2464E+01  Stress= .2289E-03
  #  5  X= .8500E+01  Y= .1670E+01  Stress= .1501E-03
  #  6  X= .7294E+01  Y= .1170E+01  Stress= .7748E-04
  #  7  X= .6000E+01  Y= .1000E+01  Stress= .1265E-06
  #  8  X= .4706E+01  Y= .1170E+01  Stress=-.7871E-04

                           .

  # 14  X= .1170E+01  Y= .7294E+01  Stress=-.2964E-03
  # 15  X= .1670E+01  Y= .8500E+01  Stress=-.2656E-03
  # 16  X= .2464E+01  Y= .9535E+01  Stress=-.2166E-03
  # 17  X= .3500E+01  Y= .1033E+02  Stress=-.1528E-03
  # 18  X= .4706E+01  Y= .1083E+02  Stress=-.7871E-04
  # 19  X= .6000E+01  Y= .1100E+02  Stress= .1265E-06
  # 20  X= .7294E+01  Y= .1083E+02  Stress= .7748E-04
  # 21  X= .8500E+01  Y= .1033E+02  Stress= .1501E-03
  # 22  X= .9535E+01  Y= .9535E+01  Stress= .2289E-03
  # 23  X= .9536E+01  Y= .8000E+01  Stress= .2888E-03
  # 24  X= .9536E+01  Y= .7000E+01  Stress= .2507E-03
```

Computed results for the absolute value of viscous stress are plotted in Figures 2-6d and 2-6e; actual stresses along the borehole/cuttings bed contour are explicitly given in Tables 2-8 and 2-9. In Figure 2-6d, the weak symmetry about the horizontal row of zeros indicates that the influence of the bed is a local one (the "zeros" actually contain unprinted fractional values). But although bed effects are local in this sense, they do affect total flow rate significantly, as is known experimentally and computed here.

Table 2-9

Example 2: Stress "AppVisc × dU(y,x)/dy" (psi)

```
Results for borehole annular boundary, Contour No. 11:
 #  1  X= .9536E+01  Y= .6000E+01  Stress= .1115E-04
 #  2  X= .9536E+01  Y= .5000E+01  Stress= .3537E-05
 #  3  X= .9536E+01  Y= .4000E+01  Stress=-.3319E-05
 #  4  X= .9535E+01  Y= .2464E+01  Stress=-.9696E-04
 #  5  X= .8500E+01  Y= .1670E+01  Stress=-.2353E-03
 #  6  X= .7294E+01  Y= .1170E+01  Stress=-.2845E-03
 #  7  X= .6000E+01  Y= .1000E+01  Stress=-.3012E-03
                         .
 # 15  X= .1670E+01  Y= .8500E+01  Stress= .1533E-03
 # 16  X= .2464E+01  Y= .9535E+01  Stress= .2166E-03
 # 17  X= .3500E+01  Y= .1033E+02  Stress= .2646E-03
 # 18  X= .4706E+01  Y= .1083E+02  Stress= .2937E-03
 # 19  X= .6000E+01  Y= .1100E+02  Stress= .3012E-03
 # 20  X= .7294E+01  Y= .1083E+02  Stress= .2845E-03
 # 21  X= .8500E+01  Y= .1033E+02  Stress= .2353E-03
 # 22  X= .9535E+01  Y= .9535E+01  Stress= .9696E-04
 # 23  X= .9536E+01  Y= .8000E+01  Stress= .3319E-05
 # 24  X= .9536E+01  Y= .7000E+01  Stress=-.3537E-05
```

Finally, Figure 2-6f displays the dissipation function as it varies in the annular cross-section. In this example, the lower part of the annulus near the cuttings bed is relatively nondissipative. Typical results are given in Table 2-10.

Table 2-10

Example 2: Dissipation Function (lbf/(sec × in.^{2}))

```
Results for pipe/collar boundary, Contour No. 1:
 #  1  X= .8000E+01  Y= .6000E+01  DissipFn= .1157E-01
 #  2  X= .7932E+01  Y= .5482E+01  DissipFn= .1249E-01
 #  3  X= .7732E+01  Y= .5000E+01  DissipFn= .1597E-01
 #  4  X= .7414E+01  Y= .4586E+01  DissipFn= .2124E-01
 #  5  X= .7000E+01  Y= .4268E+01  DissipFn= .2488E-01
                         .
 # 18  X= .5482E+01  Y= .7932E+01  DissipFn= .2813E-01
 # 19  X= .6000E+01  Y= .8000E+01  DissipFn= .2766E-01
 # 20  X= .6518E+01  Y= .7932E+01  DissipFn= .2675E-01
 # 21  X= .7000E+01  Y= .7732E+01  DissipFn= .2488E-01
 # 22  X= .7414E+01  Y= .7414E+01  DissipFn= .2124E-01
 # 23  X= .7732E+01  Y= .7000E+01  DissipFn= .1597E-01
 # 24  X= .7932E+01  Y= .6518E+01  DissipFn= .1249E-01
```

Example 3. Highly Eccentric Circular Pipe and Borehole

We now refer to the concentric geometry of Example 1, but displace the pipe downward by 2 inches (the pipe and borehole radii are 2 and 5 inches, respectively). The program allows us to add cuttings beds and general wall deformations by modifying the boundary coordinates as before. For comparative purposes, we will not do so, thus leaving the cross-sectional areas here and in Example 1 identical. The grid selected by the computer analysis is shown below in Figure 2-7a. We again assume a power law fluid with an exponent of n = 0.7240 and a consistency factor of .1861E-04 lbf sec^n/in.^{2} The axial pressure gradient is still .3890E-02 psi/ft. For brevity, we will omit tabulated results because they are similar to those in Examples 1 and 2.

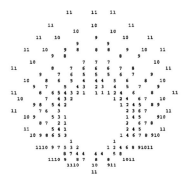

Figure 2-7a. Mesh system for eccentric circles.

The annular volume flow rate for this geometry is 883.7 gal/min, significantly higher than the 457.8 gal/min obtained for the concentric flow of Example 1. Again, this large increase is consistent with experimental observations indicating that higher eccentricity increases flow rates. The plots shown in Figures 2-7b to 2-7f for axial velocity, apparent viscosity, stress and dissipation function should be compared with earlier figures; the comparison reveals the qualitative and quantitative differences between the three annular geometries considered so far. As an additional check, we evaluated the foregoing geometry with all input parameters unchanged, except that the fluid exponent is now increased to 1.5. The fluid, becoming dilatant instead of pseudoplastic, should possess a narrower velocity profile and consequently support less volume flow. Figure 2-7g displays the axial velocity solution computed. The calculated annular volume flow rate of 149.5 gal/min is much smaller, as required, than the 883.7 gal/min computed above.

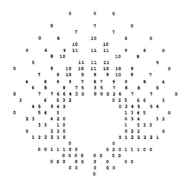

Figure 2-7b. Annular velocity.

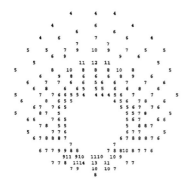

Figure 2-7c. Apparent viscosity.

Figure 2-7d. Stress "AppVisc × dU(y,x)/dx."

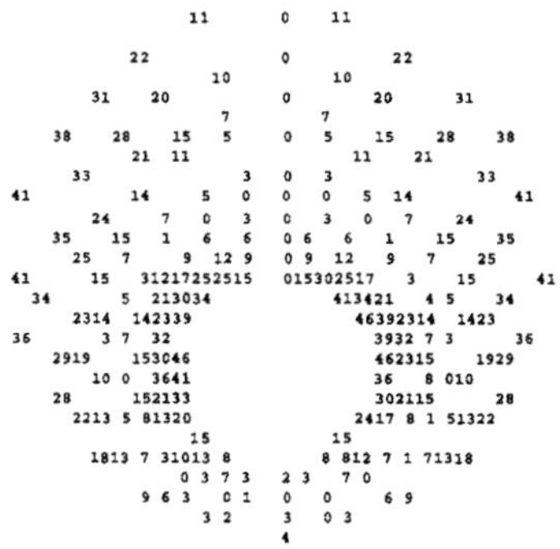

Figure 2-7e. Stress "AppVisc × dU(y,x)/dy."

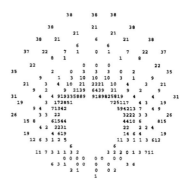

Figure 2-7f. Dissipation function.

Figure 2-7g. Annular velocity.

Example 4. Square Drill Collar in a Circular Hole

Square drill collars are sometimes used to control drillpipe sticking and dogleg severity. In this final comparative example, we consider a 5-inch-radius borehole containing a centered square drill collar having 4-inch sides. These relative dimensions are selected for plotting purposes only. In all cases, the finite difference program will accurately calculate and tabulate output quantities, even if the ASCII plotter lacks sufficient spatial resolution.

We will assume the same flow parameters as Example 1, where we had obtained peak annular velocities of 39 in./sec and a total flow rate of 457.8 gal/min. But here, tabulated results show that peak velocities of 33 in./sec are obtained near the center of the annulus; also, the total volume flow rate is 334.7 gal/min. The corresponding velocity plot is shown in Figure 2-8a.

These rates are smaller than those of Example 1 because the "4-inch square" blocks more area than its inscribed "4-inch-diameter circle." Thus, computed results are consistent with behavior expected on physical grounds. Figures 2-8b to 2-8e are given without discussion; note how the outline of the square drill collar is adequately represented throughout.

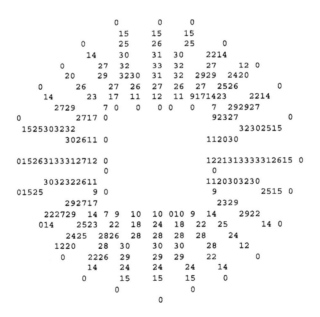

Figure 2-8a. Annular velocity, square drill collar.

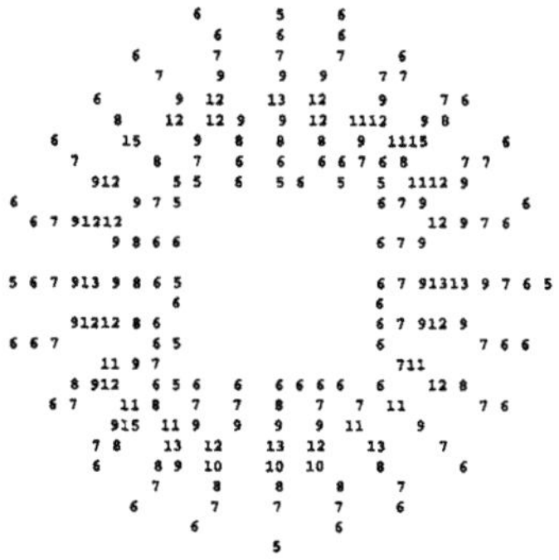

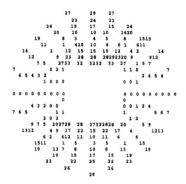

Figure 2-8b. Apparent viscosity.

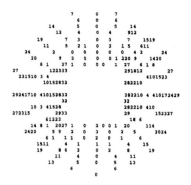

Figure 2-8d. Stress "AppVisc × dU(y,x)/dy."

Figure 2-8c. Stress "AppVisc × dU(y,x)/dx."

Figure 2-8e. Dissipation function.

The comprehensive results computed for Examples 1 to 4 summarize the first suite of computations, where fluid model, properties and pressure gradient were fixed throughout, with only the annular geometry changing from run to run. The total flow rate trends and orders of magnitude for the physical quantities predicted are consistent with known empirical observation.

Next we describe results obtained for a "small diameter hole" and a "large diameter hole," using both a power law and a Bingham plastic model. These calculations were used in planning a horizontal well, using an experimental drilling fluid developed by a mud company. Because the exact rheology was open to question, two fluid models were used to bracket the performance of the mud insofar as hole cleaning was concerned. For further developments on hole cleaning, the reader should refer to Chapter 5, which deals with applications.

In the first set of calculations (Examples 5 and 6), the annular flow in the "small diameter hole" is evaluated for a Bingham plastic and for a power law fluid, assuming a volume flow rate of 500 gpm. The computer analysis iterates to find the appropriate pressure gradient, and terminates once the sought rate falls within 1% of the target. In the second set (Examples 7 and 8), similar calculations are pursued for the "large diameter hole." In the fifth and final simulation, a cuttings bed is added to the floor of the annulus. The results importantly show that cuttings beds do affect effective rheological properties, and that they ought to be included in routine operations planning.

Example 5. Small Hole, Bingham Plastic Model

We will consider a pipe radius of 2.50 inches, and a hole radius of 4.25 inches; the pipe is displaced halfway down, by 0.90 inch. The total volume flow rate is assumed to be 500 gpm. We will determine the pressure gradient required to support this flow rate and calculate detailed flow properties. Again, these methods do not involve any geometric approximation. We solve this problem first assuming a Bingham plastic, and next a power law fluid. When the appropriate geometric parameters are entered, the computer model automatically generates a boundary conforming mesh, providing high resolution where physical gradients are large. This mesh is shown in Figure 2-9a.

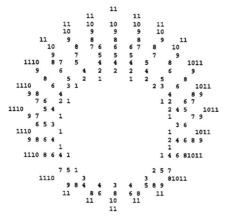

Figure 2-9a. Mesh system for small hole.

Again the actual coordinate lines are visually constructed by drawing ovals through lines of constant elevations, and then drawing orthogonals through these curves. We now take a Bingham plastic, with a plastic viscosity of 25 cp, a yield stress of 0.00139 psi, and a target flow rate of 500 gpm. The program tests different pressure gradients using a half-interval method, calculates the corresponding flow rates, and continues until the target of 500 gpm is met within 1%. These intermediate results also provide "pressure gradient vs. flow rate" information for field applications. Linearity is expected only for Newtonian flows; here, the variation is "weakly nonlinear." From the output, we have

```
O   Axial pressure gradient of .1500E-01 psi/ft
    yields volume flow rate of .2648E+03 gal/min.

O   Axial pressure gradient of .2000E-01 psi/ft
    yields volume flow rate of .3443E+03 gal/min.
```

and so on, until,

```
O   Axial pressure gradient of .3187E-01 psi/ft
    yields volume flow rate of .4980E+03 gal/min.

    Pressure gradient found iteratively, .3187E-01 psi/ft,
    yielding .4980E+03 gal/min vs target .5000E+03 gal/min.
```

At this point, the velocity solution as it depends on annular position is obtained as shown in Figure 2-9b, and numerical results are written to output files for printing. Note how the no-slip condition is identically satisfied at all solid surfaces. Also, a plug flow regime having a speed of 65 in./sec was determined to occupy most of the annular space.

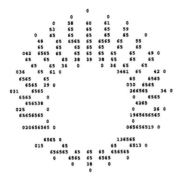

Figure 2-9b. Annular velocity, Bingham plug flow in small hole.

Figure 2-9d. Stress "AppVisc × dU(y,x)/dx."

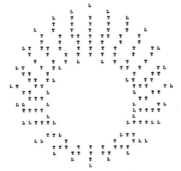

Figure 2-9c. Laminar and turbulent flow regimes.

Figure 2-9e. Shear rate dU(y,x)/dx.

Stability analyses are difficult to carry out for eccentric flows, and no claim is made to have solved the problem. However, a facility was created to provide a display of local Reynolds number, based on borehole diameter, local velocity, and apparent viscosity. These are then checked against a critical Reynolds number R_c entered by the user to produce a "laminar versus turbulent flow map." In the above, a mud with a specific gravity of 1.5 and R_c of 2,000 was assumed; the latter applies only to Newtonian flows in smooth pipes without inlet entry effects. In Figure 2-9c, the program displays "L" for laminar and "T" for turbulent flow. Reynolds numbers are printed in output files, for example,

```
Average Reynolds number, bottom half annulus = .3770E+04
Average Reynolds number, entire annulus = .3985E+04
```

Two components of viscous stress are computed. That in the vertical direction, particularly important to cuttings bed removal, is shown in Figure 2-9d. The shear rate corresponding to this stress is also obtained as part of the exact solution; its computed values (in reciprocal seconds) within the annular geometry are shown in Figure 2-9e. Note how "zeros" correctly indicate zero shear in plug-dominated regions. The annular flow model also provides areal averages of all quantities, for the "bottom half" annulus where drilled cuttings settle and for the entire annulus. The former are useful in assessing potential dangers in cuttings removal. Both averages are summarized in Table 2-11.

Table 2-11
Example 5: Summary, Average Quantities

```
TABULATION OF CALCULATED AVERAGE QUANTITIES
Area weighted means of absolute values taken over
BOTTOM HALF of annular cross-section ...
O  Average annular velocity = .4445E+02 in/sec
O  Average stress, AppVis x dU/dx, = .8316E-03 psi
O  Average stress, AppVis x dU/dy, = .7333E-03 psi
O  Average dissipation = .1045E+00 lbf/(sec sq in)
O  Average shear rate dU/dx = .3384E+02 1/sec
O  Average shear rate dU/dy = .3123E+02 1/sec

TABULATION OF CALCULATED AVERAGE QUANTITIES
Area weighted means of absolute values taken over
ENTIRE annular (y,x) cross-section ...
O  Average annular velocity = .4662E+02 in/sec
O  Average stress, AppVis x dU/dx, = .8169E-03 psi
O  Average stress, AppVis x dU/dy, = .8032E-03 psi
O  Average dissipation = .1324E+00 lbf/(sec sq in)
O  Average shear rate dU/dx = .4047E+02 1/sec
O  Average shear rate dU/dy = .4038E+02 1/sec
```

Plots of all quantities above and below the drillpipe, using convenient text plots, are available for all runs. Self-explanatory plots are shown in Figures 2-9f and 2-9g for velocities and viscous stresses. Note that the values of stress are not entirely zero inside the plug region; but they are less than yield, as required.

```
VERTICAL SYMMETRY PLANE ABOVE DRILLPIPE
Axial velocity distribution (in/sec):
    x                          0

  1.00        .0000E+00    |
  1.38        .6057E+02    |                            *
  1.72        .6591E+02    |                              *
  2.04        .6591E+02    |                              *
  2.32        .6591E+02    |                              *
  2.59        .6591E+02    |                              *
  2.84        .6591E+02    |                              *
  3.06        .6591E+02    |                              *
  3.27        .6591E+02    |                              *
  3.47        .3920E+02    |                  *
  3.65        .0000E+00    |


VERTICAL SYMMETRY PLANE BELOW DRILLPIPE
Axial velocity distribution (in/sec):
    x                          0

  8.65        .0000E+00    |
  8.70        .9162E+01    | *
  8.76        .6591E+02    |                              *
  8.83        .6591E+02    |                              *
  8.90        .6591E+02    |                              *
  8.98        .6591E+02    |                              *
  9.06        .6591E+02    |                              *
  9.16        .6591E+02    |                              *
  9.26        .6591E+02    |                              *
  9.37        .3833E+02    |                  *
  9.50        .0000E+00    |
```

Figure 2-9f. Vertical plane velocity.

```
VERTICAL SYMMETRY PLANE ABOVE DRILLPIPE
Viscous stress, AppVis x dU/dx  (psi):
    x                              0

  1.00       -.2671E-02                 |
  1.38       -.1882E-02        *        |
  1.72       -.1136E-02            *    |
  2.04       -.5755E-03               * |
  2.32       -.1058E-03                *|
  2.59        .3245E-03                |*
  2.84        .7604E-03                | *
  3.06        .1246E-02                |  *
  3.27        .1881E-02                |    *
  3.47        .2071E-02                |     *
  3.65        .1960E-02                |    *


VERTICAL SYMMETRY PLANE BELOW DRILLPIPE
Viscous stress, AppVis x dU/dx  (psi):
    x                              0

  8.65       -.2312E-02        *        |
  8.70       -.1901E-02          *      |
  8.76       -.1446E-02           *     |
  8.83       -.8806E-03              *  |
  8.90       -.4380E-03               * |
  8.98       -.1057E-04                *|
  9.06        .4153E-03                |*
  9.16        .8532E-03                | *
  9.26        .1169E-02                |  *
  9.37        .2245E-02                |     *
  9.50        .3567E-02                |       *
```

Figure 2-9g. Vertical plane viscous stress.

Example 6. Small Hole, Power Law Fluid

We repeat Example 5, but assume a power law fluid model. The power law index n and the consistency factor k can be inputted directly as in Examples 1-4. However, the program also allows users to input Fann dial readings, from which (n,k) values are internally calculated. In the present case we choose the latter option. Intermediate pressure and flow rate results are tabulated below.

```
POWER LAW FLOW OPTION SELECTED

1st Fann dial reading of .1500E+02 with corresponding
rpm of .1300E+02 assumed.

2nd Fann dial reading of .2000E+02 with corresponding
rpm of .5000E+02 assumed.  We calculate "n" and "K".

Power law fluid assumed, with exponent "n" equal
to .2136E+00 and consistency factor of .5376E-03

lbf sec^n/sq in.

Target flow rate of .5000E+03 gal/min specified.
```

Iterating on pressure gradient to match flow rate ...

```
O  Axial pressure gradient of .5000E-02 psi/ft
   yields volume flow rate of .5250E-02 gal/min.

O  Axial pressure gradient of .1000E-01 psi/ft
   yields volume flow rate of .1349E+00 gal/min.
```

until we obtain the converged result

```
O  Axial pressure gradient of .5781E-01 psi/ft
   yields volume flow rate of .5003E+03 gal/min.
```

These (large) pressure gradients and those of Example 5 are in agreement with laboratory data obtained by two different oil companies for this experimental mud. Because power law models exclude the possibility of plug flow, the velocity distribution calculated is somewhat different.

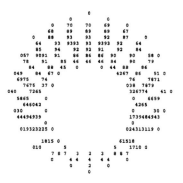

Figure 2-10a. Annular velocity, power law flow in small hole.

In Figure 2-10a, note how no-slip conditions are again satisfied, and how velocity maximums are correctly obtained near the "center" of the annulus. Figures 2-10b and 2-10c show computed stresses and shear rates. From Table 2-12 we observe that the "bottom half" average velocities are one-third of those for the entire annulus. Plots for exact annular velocities above and below the drillpipe in Figure 2-10d show even larger contrasts.

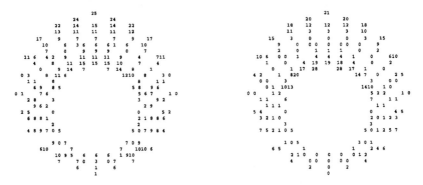

Figure 2-10b. Stress "AppVisc × dU(y,x)/dx."

Figure 2-10c. Shear rate dU(y,x)/dx.

Table 2-12
Example 6: Summary, Average Quantities

```
TABULATION OF CALCULATED AVERAGE QUANTITIES
Area weighted means of absolute values taken over
BOTTOM HALF of annular cross-section ...
o  Average annular velocity = .1360E+02 in/sec
o  Average apparent viscosity = .6028E-04 lbf sec/sq in
o  Average stress, AppVis x dU/dx, = .6395E-03 psi
o  Average stress, AppVis x dU/dy, = .6879E-03 psi
o  Average dissipation = .4688E-01 lbf/(sec sq in)
o  Average shear rate dU/dx = .2139E+02 1/sec
o  Average shear rate dU/dy = .3168E+02 1/sec
o  Average Stokes product = .6083E-03 lbf/in

TABULATION OF CALCULATED AVERAGE QUANTITIES
Area weighted means of absolute values taken over
ENTIRE annular (y,x) cross-section ...
o  Average annular velocity = .3841E+02 in/sec
o  Average apparent viscosity = .6230E-04 lbf sec/sq in
o  Average stress, AppVis x dU/dx, = .7269E-03 psi
o  Average stress, AppVis x dU/dy, = .7239E-03 psi
o  Average dissipation = .9535E-01 lbf/(sec sq in)
o  Average shear rate dU/dx = .4191E+02 1/sec
o  Average shear rate dU/dy = .4855E+02 1/sec
o  Average Stokes product = .3266E-02 lbf/in
```

```
VERTICAL SYMMETRY PLANE ABOVE DRILLPIPE
Axial velocity distribution (in/sec):
     X                    0

   1.00      .0000E+00    |
   1.38      .7068E+02    |                        *
   1.72      .8983E+02    |                          *
   2.04      .9341E+02    |                           *
   2.32      .9361E+02    |                           *
   2.59      .9353E+02    |                           *
   2.84      .9219E+02    |                           *
   3.06      .8696E+02    |                          *
   3.27      .7368E+02    |                      *
   3.47      .4686E+02    |              *
   3.65      .0000E+00    |
```

```
VERTICAL SYMMETRY PLANE BELOW DRILLPIPE
Axial velocity distribution (in/sec):
     X                    0

   8.65      .0000E+00    |
   8.70      .1642E+01    |          *
   8.76      .2797E+01    |              *
   8.83      .3560E+01    |                 *
   8.90      .4019E+01    |                   *
   8.98      .4244E+01    |                    *
   9.06      .4265E+01    |                     *
   9.16      .4036E+01    |                    *
   9.26      .3419E+01    |                 *
   9.37      .2188E+01    |           *
   9.50      .0000E+00    |
```

Figure 2-10d. Vertical plane velocity.

Example 7. Large Hole, Bingham Plastic

We next consider a 2.5-inch-radius drillpipe residing in a 6.1-inch-radius borehole. The pipe is displaced halfway down by 1.80 inches. Again, the computer code produces a boundary conforming grid system for exact calculations. This mesh is shown in Figure 2-11a.

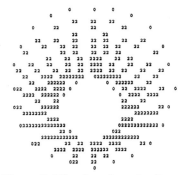

Figure 2-11a. Mesh system for large hole.

Figure 2-11b. Annular velocity, Bingham plug flow in large hole.

We will first consider a Bingham plastic fluid. The plastic viscosity is taken to be 25 cp, the yield stress as 0.00139 psi, and the target flow rate again 500 gpm. Iterating on pressure gradient to match flow rate, we find the following "pressure gradient versus flow rate" signature.

```
O  Axial pressure gradient of .5000E-02 psi/ft
   yields volume flow rate of .7026E+03 gal/min.

O  Axial pressure gradient of .3750E-02 psi/ft
   yields volume flow rate of .5270E+03 gal/min.

O  Axial pressure gradient of .3555E-02 psi/ft
   yields volume flow rate of .4995E+03 gal/min.

   Pressure gradient found iteratively, .3555E-02 psi/ft,
   yielding .4995E+03 gal/min vs target .5000E+03 gal/min.
```

Since the annular space is large compared with the "small diameter" run, the computed viscous stresses are smaller. In this particular calculation, the results indicate that they are well below yield. Hence, the entire annular flow moves as a solid plug at a constant velocity of 22 in/sec as shown in Figure 2-11b. At solid surfaces, the velocity profile rapidly adjusts to "0" to satisfy no-slip conditions, thus giving an average speed of approximately 18 in/sec.

Table 2-13
Example 7: Summary, Average Quantities

```
TABULATION OF CALCULATED AVERAGE QUANTITIES
Area weighted means of absolute values taken over
BOTTOM HALF of annular cross-section ...
O  Average annular velocity = .1870E+02 in/sec
O  Average stress, AppVis x dU/dx, = .1966E-03 psi
O  Average stress, AppVis x dU/dy, = .1656E-03 psi
O  Average dissipation = .4930E-02 lbf/(sec sq in)
O  Average shear rate dU/dx = .6637E+01 1/sec
O  Average shear rate dU/dy = .4466E+01 1/sec

TABULATION OF CALCULATED AVERAGE QUANTITIES
Area weighted means of absolute values taken over
ENTIRE annular (y,x) cross-section ...
O  Average annular velocity = .1870E+02 in/sec
O  Average stress, AppVis x dU/dx, = .1771E-03 psi
O  Average stress, AppVis x dU/dy, = .1716E-03 psi
O  Average dissipation = .4724E-02 lbf/(sec sq in)
O  Average shear rate dU/dx = .5450E+01 1/sec
O  Average shear rate dU/dy = .4794E+01 1/sec
```

Example 8. Large Hole, Power Law Fluid

Now we repeat Example 7, assuming instead a power law fluid. The results for (n,k) are summarized below, as are the simulation results.

```
POWER LAW FLOW OPTION SELECTED

1st Fann dial reading of .1500E+02 with corresponding
rpm of .1300E+02 assumed.

2nd Fann dial reading of .2000E+02 with corresponding
rpm of .5000E+02 assumed.  We calculate "n" and "k."

Power law fluid assumed, with exponent "n" equal
to .2136E+00 and consistency factor of .5376E-03

lbf sec^n/sq in.

Target flow rate of .5000E+03 gal/min specified.
```

Iterating on pressure gradient to match flow rate ...

```
O   Axial pressure gradient of .5000E-02 psi/ft
    yields volume flow rate of .6528E+00 gal/min.

O   Axial pressure gradient of .1000E-01 psi/ft
    yields volume flow rate of .1593E+02 gal/min.
```

and so on, until,

```
O   Axial pressure gradient of .2094E-01 psi/ft
    yields volume flow rate of .5039E+03 gal/min.
```

The computed velocity field shown in Figure 2-12a is somewhat more interesting than the plug flow obtained in Figure 2-11b. Other calculated quantities are given in Figures 2-12b to 2-12g without explanation. Relevant areal averages are listed in Table 2-14.

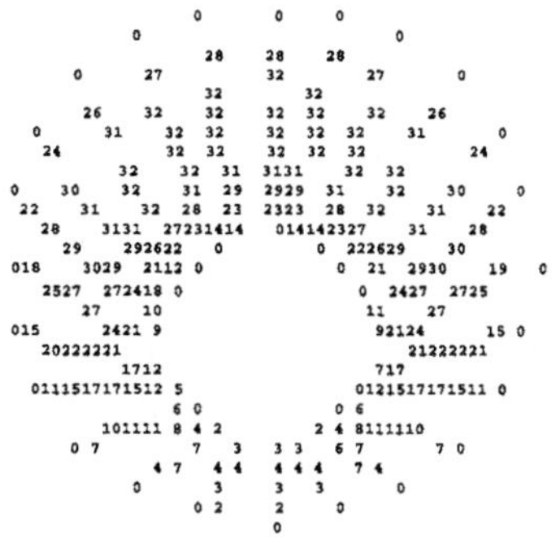

Figure 2-12a. Annular velocity, power law flow in large hole.

Table 2-14

Example 8: Summary, Average Quantities

TABULATION OF CALCULATED AVERAGE QUANTITIES
Area weighted means of absolute values taken over
BOTTOM HALF of annular cross-section ...
o Average annular velocity = .7696E+01 in/sec
o Average apparent viscosity = .1198E-03 lbf sec/sq in
o Average stress, AppVis x dU/dx, = .5251E-03 psi
o Average stress, AppVis x dU/dy, = .5130E-03 psi
o Average dissipation = .1034E-01 lbf/(sec sq in)
o Average shear rate dU/dx = .6954E+01 1/sec
o Average shear rate dU/dy = .8676E+01 1/sec
o Average Stokes product = .9454E-03 lbf/in

TABULATION OF CALCULATED AVERAGE QUANTITIES
Area weighted means of absolute values taken over
ENTIRE annular (y,x) cross-section ...
o Average annular velocity = .1527E+02 in/sec
o Average apparent viscosity = .2074E-03 lbf sec/sq in
o Average stress, AppVis x dU/dx, = .5562E-03 psi
o Average stress, AppVis x dU/dy, = .5272E-03 psi
o Average dissipation = .1530E-01 lbf/(sec sq in)
o Average shear rate dU/dx = .9280E+01 1/sec
o Average shear rate dU/dy = .1056E+02 1/sec
o Average Stokes product = .5144E-02 lbf/in

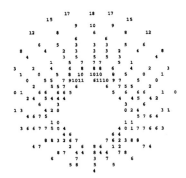

Figure 2-12b. Stress "AppVisc ×
dU(y,x)/dx."

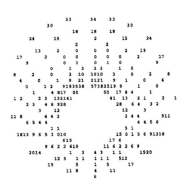

Figure 2-12c. Shear rate
dU(y,x)/dx.

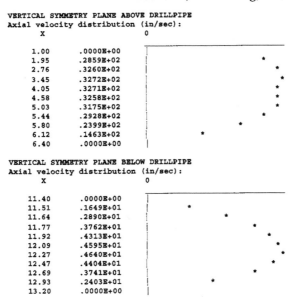

```
VERTICAL SYMMETRY PLANE ABOVE DRILLPIPE
Axial velocity distribution (in/sec):
    X                    0

   1.00        .0000E+00    |
   1.95        .2859E+02    |                              *
   2.76        .3260E+02    |                                *
   3.45        .3272E+02    |                                 *
   4.05        .3271E+02    |                                 *
   4.58        .3258E+02    |                                 *
   5.03        .3175E+02    |                                *
   5.44        .2928E+02    |                              *
   5.80        .2399E+02    |                        *
   6.12        .1463E+02    |               *
   6.40        .0000E+00    |

VERTICAL SYMMETRY PLANE BELOW DRILLPIPE
Axial velocity distribution (in/sec):
    X                    0

  11.40        .0000E+00    |
  11.51        .1649E+01    |       *
  11.64        .2890E+01    |            *
  11.77        .3762E+01    |                *
  11.92        .4313E+01    |                  *
  12.09        .4595E+01    |                    *
  12.27        .4640E+01    |                     *
  12.47        .4404E+01    |                   *
  12.69        .3741E+01    |                *
  12.93        .2403E+01    |          *
  13.20        .0000E+00    |
```

Figure 2-12d. Vertical plane velocity.

```
VERTICAL SYMMETRY PLANE ABOVE DRILLPIPE
Apparent viscosity distribution (lbf sec/sq in):
    X                    0

   1.00        .5410E-04    |
   1.95        .5410E-04    |
   2.76        .2425E-03    *
   3.45        .3657E-02    |                              *
   4.05        .2792E-02    |                       *
   4.58        .5468E-03    |  *
   5.03        .1873E-03    *
   5.44        .8667E-04    |
   5.80        .4789E-04    |
   6.12        .2969E-04    |
   6.40        .1149E-04    |

VERTICAL SYMMETRY PLANE BELOW DRILLPIPE
Apparent viscosity distribution (lbf sec/sq in):
    X                    0

  11.40        .6464E-04    | *
  11.51        .1046E-03    |   *
  11.64        .1446E-03    |     *
  11.77        .2176E-03    |        *
  11.92        .3630E-03    |             *
  12.09        .5925E-03    |                    *
  12.27        .5426E-03    |                  *
  12.47        .3238E-03    |            *
  12.69        .1997E-03    |       *
  12.93        .1332E-03    |    *
  13.20        .6672E-04    | *
```

Figure 2-12e. Vertical plane apparent viscosity.

```
VERTICAL SYMMETRY PLANE ABOVE DRILLPIPE
Viscous stress, AppVis x dU/dx  (psi):
    X                                                0

   1.00      -.1857E-02      |                        |
   1.95      -.1003E-02           *                   |
   2.76      -.6673E-03               *               |
   3.45      -.3194E-03                   *           |
   4.05       .3437E-03                      *        |
   4.58       .5351E-03                        *      |
   5.03       .7158E-03                        *      |
   5.44       .8824E-03                          *    |
   5.80       .1037E-02                           *   |
   6.12       .1180E-02                            *  |
   6.40       .6649E-03                         *     |


VERTICAL SYMMETRY PLANE BELOW DRILLPIPE
Viscous stress, AppVis x dU/dx  (psi):
    X                                                0

  11.40      -.6990E-03      *                        |
  11.51      -.8801E-03                               |
  11.64      -.8693E-03                               |
  11.77      -.8743E-03                               |
  11.92      -.9128E-03                               |
  12.09      -.8764E-03                               |
  12.27      -.3433E-03                  *            |
  12.47       .1435E-03                     *         |
  12.69       .3946E-03                        *      |
  12.93       .5310E-03                          *    |
  13.20       .4001E-03                        *      |
```

Figure 2-12f. Vertical plane viscous stress.

```
VERTICAL SYMMETRY PLANE ABOVE DRILLPIPE
Shear rate dU/dx  (1/sec):
    X                                                0

   1.00      -.3433E+02      *                        |
   1.95      -.1854E+02           *                   |
   2.76      -.2752E+01                            *  |
   3.45      -.8735E-01                             * |
   4.05       .1231E+00                               |
   4.58       .9787E+00                               |
   5.03       .3821E+01                               |
   5.44       .1018E+02                            *  |
   5.80       .2165E+02                              * |
   6.12       .3976E+02                                  *
   6.40       .5787E+02                                     *


VERTICAL SYMMETRY PLANE BELOW DRILLPIPE
Shear rate dU/dx  (1/sec):
    X                                                0

  11.40      -.1081E+02      |                        |
  11.51      -.8412E+01         *                     |
  11.64      -.6012E+01             *                 |
  11.77      -.4018E+01                *              |
  11.92      -.2515E+01                   *           |
  12.09      -.1479E+01                     *         |
  12.27      -.6328E+00                        *      |
  12.47       .4433E+00                               |
  12.69       .1976E+01                          *    |
  12.93       .3986E+01                            *  |
  13.20       .5997E+01                               *
```

Figure 2-12g. Vertical plane shear rate.

Example 9. Large Hole with Cuttings Bed

Finally, the geometry of Example 8 was reevaluated to examine the effect of a flat cuttings bed, using the pressure gradient obtained in the above run. The boundary conforming mesh generated for this annulus resolves the flowfield at the bed in sufficient detail, and provides as many "radial" meshes at the bottom as there are on top. The computed mesh is shown in Figure 2-13a. Figures 2-13b to 2-13f, and Table 2-15, summarize the self-explanatory simulation.

Table 2-15
Example 9: Summary, Average Quantities

```
TABULATION OF CALCULATED AVERAGE QUANTITIES
Area weighted means of absolute values taken over
BOTTOM HALF of annular cross-section ...
O  Average annular velocity = .6708E+01 in/sec
O  Average apparent viscosity = .1258E-03 lbf sec/sq in
O  Average stress, AppVis x dU/dx, = .4717E-03 psi
O  Average stress, AppVis x dU/dy, = .5651E-03 psi
O  Average dissipation = .9985E-02 lbf/(sec sq in)
O  Average shear rate dU/dx = .6188E+01 1/sec
O  Average shear rate dU/dy = .8557E+01 1/sec
O  Average Stokes product = .6989E-03 lbf/in

TABULATION OF CALCULATED AVERAGE QUANTITIES
Area weighted means of absolute values taken over
ENTIRE annular (y,x) cross-section ...
O  Average annular velocity = .1483E+02 in/sec
O  Average apparent viscosity = .2098E-03 lbf sec/sq in
O  Average stress, AppVis x dU/dx, = .5320E-03 psi
O  Average stress, AppVis x dU/dy, = .5513E-03 psi
O  Average dissipation = .1517E-01 lbf/(sec sq in)
O  Average shear rate dU/dx = .8943E+01 1/sec
O  Average shear rate dU/dy = .1052E+02 1/sec
O  Average Stokes product = .5027E-02 lbf/in


POWER LAW FLOW OPTION SELECTED

1st Fann dial reading of .1500E+02 with corresponding
rpm of .1300E+02 assumed.

2nd Fann dial reading of .2000E+02 with corresponding
rpm of .5000E+02 assumed.    We calculate "n" and "k"

Power law fluid assumed, with exponent "n" equal
to .2136E+00 and consistency factor of .5376E-03
lbf sec^n/sq in.
```

Axial pressure gradient assumed as .2094E-01 psi/ft.

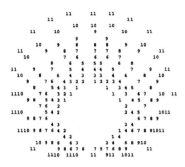

Figure 2-13a. Mesh system, large hole with cuttings bed.

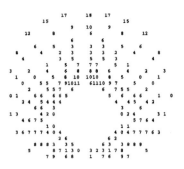

Figure 2-13c. Stress "AppVisc × dU(y,x)/dx."

Figure 2-13b. Annular velocity.

Figure 2-13d. Shear rate dU(y,x)/dx.

We have presented in this chapter, a self-contained mathematical model describing the flow of Newtonian, power law, Bingham plastic, and Herschel-Bulkley fluids through eccentric annular spaces. Again, the borehole contour need not be circular; in fact, it may be modified "point by point" to simulate cuttings beds and general wall deformations due to erosion and swelling. Likewise, the default "pipe contour" while circular, may be altered to model drillpipes and collars with or without stabilizers and centralizers.

A fast, second-order accurate, unconditionally stable finite difference scheme was used to solve the nonlinear governing PDEs. The formulation uses exact boundary conforming grid systems that eliminate the need for unrealistic simplifying assumptions about the annular geometry. "Slot flow" and "hydraulic radii" approximations are never used.

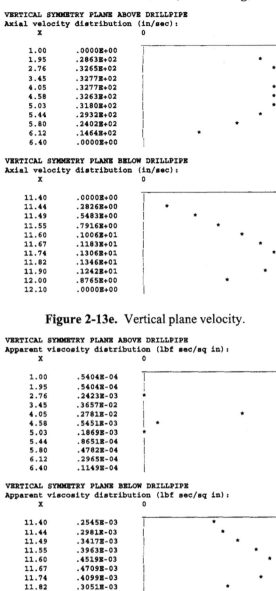

```
VERTICAL SYMMETRY PLANE ABOVE DRILLPIPE
Axial velocity distribution (in/sec):
     X                           0

   1.00       .0000E+00      |
   1.95       .2863E+02      |                        *
   2.76       .3265E+02      |                           *
   3.45       .3277E+02      |                            *
   4.05       .3277E+02      |                            *
   4.58       .3263E+02      |                           *
   5.03       .3180E+02      |                          *
   5.44       .2932E+02      |                       *
   5.80       .2402E+02      |                 *
   6.12       .1464E+02      |          *
   6.40       .0000E+00      |

VERTICAL SYMMETRY PLANE BELOW DRILLPIPE
Axial velocity distribution (in/sec):
     X                           0

  11.40       .0000E+00      |
  11.44       .2826E+00      |    *
  11.49       .5483E+00      |       *
  11.55       .7916E+00      |          *
  11.60       .1006E+01      |             *
  11.67       .1183E+01      |                *
  11.74       .1306E+01      |                  *
  11.82       .1346E+01      |                   *
  11.90       .1242E+01      |                  *
  12.00       .8765E+00      |            *
  12.10       .0000E+00      |
```

Figure 2-13e. Vertical plane velocity.

```
VERTICAL SYMMETRY PLANE ABOVE DRILLPIPE
Apparent viscosity distribution (lbf sec/sq in):
     X                           0

   1.00       .5404E-04      |
   1.95       .5404E-04      |
   2.76       .2423E-03      *
   3.45       .3657E-02      |                              *
   4.05       .2781E-02      |                         *
   4.58       .5451E-03      |   *
   5.03       .1869E-03      *
   5.44       .8651E-04      |
   5.80       .4782E-04      |
   6.12       .2965E-04      |
   6.40       .1149E-04      |

VERTICAL SYMMETRY PLANE BELOW DRILLPIPE
Apparent viscosity distribution (lbf sec/sq in):
     X                           0

  11.40       .2545E-03      |                  *
  11.44       .2981E-03      |                    *
  11.49       .3417E-03      |                      *
  11.55       .3963E-03      |                        *
  11.60       .4519E-03      |                          *
  11.67       .4709E-03      |                           *
  11.74       .4099E-03      |                        *
  11.82       .3051E-03      |                  *
  11.90       .2146E-03      |            *
  12.00       .1496E-03      |        *
  12.10       .8459E-04      |    *
```

Figure 2-13f. Vertical plane apparent viscosity.

Computed quantities include detailed plots and tables for annular velocity, apparent viscosity, viscous stress, shear rate, Stokes product, and dissipation function. To facilitate visual correlation with annular position, a portable, Fortran-based, ASCII text plotter was developed to overlay computed quantities directly on the annulus. Software displaying "vertical plane plots" was also written to enhance the interpretation of computed quantities. Sophisticated color plotting capabilities are also available, which are described later.

Calculations for several non-Newtonian flows using numerous complicated annular geometries were performed. The results, which agree with empirical observation, were computed in a stable manner in all cases. The cross-sectional displays show an unusual amount of information that is easily interpreted and understood by petroleum engineers. Moreover, the computer algorithm is fast and requires minimal hardware and graphical software investment.

REFERENCES

Bird, R.B., Stewart, W.E., and Lightfoot, E.N., *Transport Phenomena*, John Wiley & Sons, New York, 1960.

Crochet, M.J., Davies, A.R., and Walters, K., *Numerical Simulation of Non-Newtonian Flow*, Elsevier Science Publishers B.V., Amsterdam, 1984.

Lapidus, L., and Pinder, G., *Numerical Solution of Partial Differential Equations in Science and Engineering*, John Wiley & Sons, New York, 1982.

Schlichting, H., *Boundary Layer Theory*, McGraw-Hill, New York, 1968.

Slattery, J.C., *Momentum, Energy, and Mass Transfer in Continua*, Robert E. Krieger Publishing Company, New York, 1981.

Streeter, V.L., *Handbook of Fluid Dynamics*, McGraw-Hill, New York, 1961.

Thompson, J.F., Warsi, Z.U.A., and Mastin, C.W., *Numerical Grid Generation*, Elsevier Science Publishing, New York, 1985.

3

Concentric, Rotating, Annular Flow

Analytical solutions for the nonlinearly coupled axial and circumferential velocities, their deformation, stress and pressure fields, are obtained for the annular flow in an inclined borehole with a centered, rotating drillstring or casing. The closed form solutions are used to derive formulas for volume flow rate, maximum borehole wall stress, apparent viscosity, and other quantities as functions of "r." The analysis is restricted to Newtonian and power law fluids. Our Newtonian results are *exact* solutions to the viscous Navier-Stokes equations without geometric approximation. For power law fluids, the analytical results assume a *narrow annulus*, but reduce to the Newtonian solutions in the "n=1" limit. All solutions satisfy no-slip viscous boundary conditions at both the rotating drillstring and the borehole wall. The formulas are also explicit; they require no iteration and are easily programmed on pocket calculators. Extensive analytical and calculated results are given, which elucidate the physical differences between the two fluid types.

GENERAL GOVERNING EQUATIONS

The equations governing general fluid motion are available from many excellent textbooks on continuum mechanics (Schlichting, 1968; Slattery, 1981). We will cite these equations without proof. Let v_r, v_θ and v_z denote Eulerian fluid velocities, and F_r, F_θ and F_z the body forces, in the r, θ and z directions, respectively. Here (r, θ,z) are standard circular cylindrical coordinates.

Also, let ρ be the *constant* fluid density and p be the pressure; and denote by S_{rr}, $S_{r\theta}$, $S_{\theta\theta}$, S_{rz}, $S_{\theta r}$, $S_{\theta z}$, S_{zr}, $S_{z\theta}$ and S_{zz} the nine elements of the general extra stress tensor $\underline{S}$. If t is time, and ∂'s represent partial derivatives, the complete equations obtained from Newton's law and mass conservation are

Momentum equation in r:

$$\rho \left(\partial v_r/\partial t + v_r \partial v_r/\partial r + v_\theta/r \; \partial v_r/\partial \theta - v_\theta^2/r + v_z \; \partial v_r/\partial z \right) = \qquad (3\text{-}1)$$

$$= F_r - \partial p/\partial r + 1/r \; \partial(rS_{rr})/\partial r + 1/r \; \partial(S_{r\theta})/\partial \theta + \partial(S_{rz})/\partial z - S_{\theta\theta}/r$$

Momentum equation in θ:

$$\rho \left(\partial v_\theta/\partial t + v_r \; \partial v_\theta/\partial r + v_\theta/r \; \partial v_\theta/\partial \theta + v_r v_\theta/r + v_z \; \partial v_\theta/\partial z \right) = \quad (3\text{-}2)$$

$$= F_\theta - 1/r \; \partial p/\partial \theta + 1/r^2 \; \partial(r^2 S_{\theta r})/\partial r + 1/r \; \partial(S_{\theta\theta})/\partial \theta + \partial(S_{\theta z})/\partial z$$

Momentum equation in z:

$$\rho \left(\partial v_z/\partial t + v_r \; \partial v_z/\partial r + v_\theta/r \; \partial v_z/\partial \theta + v_z \partial v_z/\partial z \right) = \qquad (3\text{-}3)$$

$$= F_z - \partial p/\partial z + 1/r \; \partial(rS_{z\,r})/\partial r + 1/r \; \partial(S_{z\,\theta})/\partial \theta + \partial(S_{zz})/\partial z$$

Mass continuity equation:

$$1/r \; \partial(rv_r)/\partial r + 1/r \; \partial v_\theta/\partial \theta + \partial v_z/\partial z = 0 \qquad (3\text{-}4)$$

These equations apply to all Newtonian and non-Newtonian fluids. In continuum mechanics, the most common class of empirical models for isotropic, incompressible fluids assumes that $\underline{S}$ can be related to the rate of deformation tensor $\underline{D}$ by a relationship of the form

$$\underline{S} = 2\,N(\Gamma)\,\underline{D} \qquad (3\text{-}5)$$

where the elements of $\underline{D}$ are

$$D_{rr} = \partial v_r/\partial r \qquad (3\text{-}6)$$

$$D_{\theta\theta} = 1/r \; \partial v_\theta/\partial \theta + v_r/r \qquad (3\text{-}7)$$

$$D_{zz} = \partial v_z/\partial z \qquad (3\text{-}8)$$

$$D_{r\theta} = D_{\theta r} = [r \; \partial(v_\theta/r)/\partial r + 1/r \; \partial v_r/\partial \theta]\,/2 \qquad (3\text{-}9)$$

$$D_{rz} = D_{z\,r} = [\partial v_r/\partial z + \partial v_z/\partial r]\,/2 \qquad (3\text{-}10)$$

$$D_{\theta z} = D_{z\theta} = [\partial v_\theta/\partial z + 1/r \; \partial v_z/\partial \theta]\,/2 \qquad (3\text{-}11)$$

In Equation 3-5, $N(\Gamma)$ is the "apparent viscosity function" satisfying

$$N(\Gamma) > 0 \qquad (3\text{-}12)$$

$\Gamma(r,\theta,z)$ being the scalar functional of v_r, v_θ and v_z defined by the tensor operation

$$\Gamma = \{\,2 \; \text{trace}\,(\underline{D}\bullet\underline{D})\,\}^{1/2} \qquad (3\text{-}13)$$

These considerations are still very general. Let us examine an important and practical simplification. The Ostwald-de Waele model for two-parameter "power law fluids" assumes that the apparent viscosity satisfies

$$N(\Gamma) = k\, \Gamma^{n-1} \tag{3-14}$$

where the fluid exponent "n" and the consistency factor "k" are constants. Power law fluids are "pseudoplastic" when $0 < n < 1$, Newtonian when $n = 1$, and "dilatant" when $n > 1$. Most drilling fluids are pseudoplastic. In the limit ($n=1$, $k=\mu$), Equation 3-14 reduces to a Newtonian model with $N(\Gamma) = \mu$, where μ is the laminar viscosity; here, stress is linearly proportional to shear rate.

On the other hand, when n and k take on general values, the apparent viscosity function becomes somewhat complicated. For isotropic, rotating flows without velocity dependence on the azimuthal coordinate θ, the function Γ in Equation 3-14 takes the form

$$\Gamma = [\, (\partial v_z/\partial r)^2 + r^2\, (\partial\{v_\theta/r\}/\partial r)^2\,]^{1/2} \tag{3-15}$$

as we will show, so that Equation 3-14 becomes

$$N(\Gamma) = k\, [(\partial v_z/\partial r)^2 + r^2\, (\partial\{v_\theta/r\}/\partial r)^2\,]^{(n-1)/2} \tag{3-16}$$

This apparent viscosity reduces to the conventional $N(\Gamma) = k\, (\partial v_z/\partial r)^{(n-1)}$ for "axial only" flows without rotation; and, to $N(\Gamma) = k\, (r\, \partial\{v_\theta/r\}/\partial r)^{(n-1)}$ for "rotation only" viscometer flows without axial velocity. When both axial and circumferential velocities are present, as in annular flows with drillstring rotation, neither of these simplifications applies. This leads to mathematical difficulty. Even though "v_θ (max)" is known from the rotation rate, the magnitude of the nondimensional "v_θ (max)/v_z(max)" ratio cannot be accurately estimated because v_z is highly sensitive to n, k, and pressure drop. Thus, it is impossible to determine beforehand whether or not rotation effects will be weak; simple "axial flow only" formulas cannot be used *a priori*.

Our result for Newtonian flow, an *exact* solution to the Navier-Stokes equations, is considered first, *without* geometric approximation. Then an approximate solution for pseudoplastic and dilatant power law fluids is developed for more general n's; we will derive closed form results for rotating flows using Equation 3-16 in its entirety, assuming a narrow annulus, although no further simplifications are taken. Because the mathematical manipulations are complicated, the Newtonian limit is examined first to gain insight into the general case. This is instructive because it allows us to highlight the physical differences between Newtonian and power law flows.

The annular geometry is shown in Figure 2-1. A drillpipe (or casing) and borehole combination is inclined at an angle α relative to the ground, with $\alpha = 0^0$ for horizontal and $\alpha = 90^0$ for vertical wells. "Z" denotes any point within the drillpipe or annular fluid; Section "AA" is cut normal to the local z axis.

Figure 2-2 resolves the vertical body force at "Z" due to gravity into components parallel and perpendicular to the axis. Figure 3-1 further breaks the latter into vectors in the radial and azimuthal directions of the cylindrical coordinate system at Section "AA." Physical assumptions about the drillstring and borehole flow in these coordinates are developed next. Their engineering and mathematical consistency will be evaluated, and applications formulas and detailed calculations will be given.

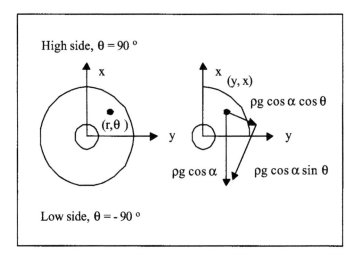

Figure 3-1.   Free body diagram, gravity in (r,θ,z) coordinates.

EXACT SOLUTION FOR NEWTONIAN FLOWS

For Newtonian flows, the stress is linearly proportional to the shear rate; the proportionality constant is the laminar fluid viscosity μ. We assume for simplicity that μ is constant. In high shear gradient flows with non-negligible heat generation, μ would depend on temperature, which in turn affects on velocity; in this event, an additional coupled energy equation would be needed. For the problem at hand, Equations 3-1 to 3-3 become

Momentum equation in r: (3-17)

$$\rho\{\partial v_r/\partial t + v_r\,\partial v_r/\partial r + v_\theta/r\;\partial v_r/\partial\theta - v_\theta^2/r + v_z\partial v_r/\partial z\} = F_r - \partial p/\partial r$$
$$+ \mu\{\partial^2 v_r/\partial r^2 + 1/r\;\partial v_r/\partial r - v_r/r^2 + 1/r^2\;\partial^2 v_r/\partial\theta^2 - 2/r^2\;\partial v_\theta/\partial\theta + \partial^2 v_r/\partial z^2\}$$

Momentum equation in θ: (3-18)

$$\rho\{\partial v_\theta/\partial t + v_r\,\partial v_\theta/\partial r + v_\theta/r\;\partial v_\theta/\partial\theta + v_r v_\theta/r + v_z\partial v_\theta/\partial z\} = F_\theta - 1/r\;\partial p/\partial\theta$$
$$+ \mu\{\partial^2 v_\theta/\partial r^2 + 1/r\;\partial v_\theta/\partial r - v_\theta/r^2 + 1/r^2\;\partial^2 v_\theta/\partial\theta^2 + 2/r^2\;\partial v_r/\partial\theta + \partial^2 v_\theta/\partial z^2\}$$

Momentum equation in z: (3-19)

$$\rho\{\partial v_z/\partial t + v_r\,\partial v_z/\partial r + v_\theta/r\;\partial v_z/\partial\theta + v_z\partial v_z/\partial z\} = F_z - \partial p/\partial z$$
$$+ \mu\{\partial^2 v_z/\partial r^2 + 1/r\;\partial v_z/\partial r + 1/r^2\;\partial^2 v_z/\partial\theta^2 + \partial^2 v_z/\partial z^2\}$$

In this section, it is convenient to rewrite Equation 3-4 in the expanded form

Mass continuity equation:

$$\partial v_r/\partial r + v_r/r + 1/r\;\partial v_\theta/\partial\theta + \partial v_z/\partial z = 0 \qquad (3\text{-}20)$$

Now consider the free body diagrams in Figures 2-1, 2-2 and 3-1. Figure 2-1 shows a straight borehole with a centered, rotating drillstring inclined at an angle α relative to the ground. Figure 2-2, referring to this geometry, resolves the gravity vector **g** into components parallel and perpendicular to the hole axis. Figure 3-1 applies to the circular cross-section AA in Figure 2-1 and introduces local cylindrical coordinates (r, θ). The "low side, $\theta = -90^o$" marks the position where cuttings beds would normally form. The force $\rho g \cos\alpha$ of Figure 2-2 is resolved into orthogonal components $\rho g \cos\alpha \sin\theta$ and $\rho\,g \cos\alpha \cos\theta$.

Physical assumptions about the flow are now given. First, it is expected that at any section AA along the borehole axis z, the velocity fields will appear to be the same; they are invariant, so z derivatives of v_r, v_θ and v_z vanish. Also, since the drillpipe and borehole walls are assumed to be impermeable, $v_r = 0$ throughout (in formation invasion modeling, this would not apply). While we do have pipe rotation, the use of circular cylindrical coordinates (with constant v_θ at the drillstring) renders the mathematical formulation steady. Thus, all time derivatives vanish. These assumptions imply that

$$- \rho v_\theta^2/r = F_r - \partial p/\partial r - 2\mu/r^2\;\partial v_\theta/\partial\theta \qquad (3\text{-}21)$$
$$\rho v_\theta/r\;\partial v_\theta/\partial\theta = F_\theta - 1/r\;\partial p/\partial\theta \qquad (3\text{-}22)$$
$$+ \mu\{\partial^2 v_\theta/\partial r^2 + 1/r\;\partial v_\theta/\partial r - v_\theta/r^2 + 1/r^2\;\partial^2 v_\theta/\partial\theta^2\}$$
$$\rho v_\theta/r\;\partial v_z/\partial\theta = F_z - \partial p/\partial z \qquad (3\text{-}23)$$
$$+ \mu\{\partial^2 v_z/\partial r^2 + 1/r\;\partial v_z/\partial r + 1/r^2\;\partial^2 v_z/\partial\theta^2\}$$
$$\partial v_\theta/\partial\theta = 0 \qquad (3\text{-}24)$$

Equation 3-24 is useful in simplifying Equations 3-21 to 3-23 further. We straightforwardly obtain

$$\rho \, v_\theta^2/r = \rho \, g \cos \alpha \sin \theta + \partial p/\partial r \tag{3-25}$$

$$0 = -\rho \, g \cos \alpha \cos \theta - 1/r \, \partial p/\partial \theta \tag{3-26}$$
$$+ \mu \{\partial^2 v_\theta/\partial r^2 + 1/r \, \partial v_\theta/\partial r - v_\theta/r^2\}$$

$$\rho \, v_\theta/r \, \partial v_z/\partial \theta = \rho \, g \sin \alpha - \partial p/\partial z \tag{3-27}$$
$$+ \mu \{\partial^2 v_z/\partial r^2 + 1/r \, \partial v_z/\partial r + 1/r^2 \, \partial^2 v_z/\partial \theta^2\}$$

where we have substituted the body force components of Figures 2-2 and 3-1. Now, since Equation 3-27 does not explicitly contain θ, it follows that v_z is independent of θ. Since we had already shown that there is no z dependence, we find $v_z = v_z$ (r) is a function of r only. Equation 3-27 therefore becomes

$$0 = \rho \, g \sin \alpha - \partial p/\partial z + \mu \{\partial^2 v_z/\partial r^2 + 1/r \, \partial v_z/\partial r\} \tag{3-28}$$

To achieve further simplicity, we resolve (without loss of generality) the pressure $p(r,\theta,z)$ into its component dynamic pressures $P(z)$ and $P^*(r)$, and its hydrostatic contribution, through the separation of variables

$$p(r,\theta,z) = P(z) + P^*(r) + z\rho g \sin \alpha - r \, \rho g \cos \alpha \sin \theta \tag{3-29}$$

This reduces the governing Navier-Stokes equations to the simpler but mathematically equivalent system

$$\partial^2 v_z/\partial r^2 + 1/r \, \partial v_z/\partial r = 1/\mu \, dP(z)/dz = \text{constant} \tag{3-30}$$

$$\partial^2 v_\theta/\partial r^2 + 1/r \, \partial v_\theta/\partial r - v_\theta/r^2 = 0 \tag{3-31}$$

$$\rho \, v_\theta^2/r = dP^*(r)/dr \tag{3-32}$$

The separation of variables introduced in Equation 3-29 and the explicit elimination of "g" in Equations 3-30 to 3-32 do not mean that gravity is unimportant; the effects of gravity are simply tracked in the $dP(z)/dz$ term of Equation 3-30. The function $P^*(r)$ will depend on the velocity solution to be obtained. Equations 3-30 to 3-32 are also significant in another respect. The velocity fields $v_z(r)$ and $v_\theta(r)$ can be obtained *independently* of each other, despite the nonlinearity of the Navier-Stokes equations, because Equations 3-30 and 3-31 physically uncouple. This decoupling occurs because the nonlinear convective terms in the original momentum equations identically vanish. Equation 3-32 is only applied (after the fact) to calculate the radial pressure field $P^*(r)$ for use in Equation 3-29. This decoupling applies only to Newtonian flows. For non-Newtonian flows, $v_z(r)$ and $v_\theta(r)$ are strongly coupled mathematically, and different solution strategies are needed.

This degeneracy with Newtonian flows means that their physical properties will be completely different from those of power law fluids. For Newtonian flows, changes in rotation rate will not affect properties in the axial direction, in contrast to non-Newtonian flows. Cuttings transport recommendations deduced, for example, using water as the working medium, cannot be extrapolated to general drilling fluids having fractional values of n, using any form of dimensional analysis. Similarly, observations for power law fluids need not apply to water. This uncoupling was, apparently, first observed by Savins and Wallick (1966), and the author is indebted to J. Savins for bringing this earlier result to his attention. Savins and Wallick noted that in Newtonian flows, no coupling among the discharge rate, axial pressure gradient, relative motion and torque through viscosity exists. But we emphasize that the coupling between v_z and v_θ reappears in eccentric geometries even for Newtonian flows.

Because Equations 3-30 and 3-31 are linear, it is possible to solve for the complete flowfield using exact classical methods. We will give all required solutions without proof, since they can be verified by direct substitution. For the *inside* of the drillpipe, the axial flow solution to Equation 3-30 satisfying no-slip conditions at the pipe radius $r = R_P$ and zero shear stress at the centerline $r = 0$ is

$$v_z(r) = (r^2 - R_P^2)/4\mu \ dP(z)/dz \qquad (3\text{-}33)$$

Similarly, the rotating flow solution to Equation 3-31 satisfying bounded flow at $r = 0$ and $v_\theta/r = \omega$ at $r = R_P$ is

$$v_\theta = \omega r \qquad (3\text{-}34)$$

This is just the expected equation for solid body rotation. Here, "ω" is a constant drillstring rotation rate. These velocity results, again, can be linearly superposed despite the nonlinearity of the underlying equations.

Now, let L denote the length of the pipe, P_{mp} be the constant pressure at the "mudpump" $z = 0$, and P^- be the drillpipe pressure at $z = L$ just upstream of the bit nozzles. Direct integration of Equation 3-32 and substitution in Equation 3-29 yield the complementary solution for pressure

$$p(r,\theta,z) = P_{mp} + (P^- - P_{mp}) \ z/L + \rho\omega^2 r^2/2$$
$$+ \rho g(z \sin \alpha - r \cos \alpha \sin \theta) + \text{constant} \qquad (3\text{-}35)$$

For the annular region between the rotating drillstring and the stationary borehole wall, the solution of Equation 3-30 satisfying no-slip conditions at the pipe radius $r = R_P$ and at the borehole radius $r = R_B$ is

$$v_z(r) = \{r^2 - R_P^2 + (R_B^2 - R_P^2) (\log r/R_P)/\log R_P/R_B\} \ 1/4\mu \ dP(z)/dz$$
$$\qquad (3\text{-}36)$$

where "log" denotes the natural logarithm. The solution of Equation 3-31 satisfying $v_\theta = 0$ at $r = R_B$ and $v_\theta = \omega r$ at $r = R_P$ is

$$v_\theta(r) = \omega R_P (R_B/r - r/R_B)/(R_B/R_P - R_P/R_B) \qquad (3-37)$$

Now let P^+ be the pressure at $z = L$ just outside of the bit nozzles, and P_{ex} be the surface exit pressure at $z = 0$. The solution for pressure from Equation 3-32 is

$$p(r,\theta,z) = P^+ + (P_{ex} - P^+)(L - z)/L + \rho g(z \sin \alpha - r \cos \alpha \sin \theta)$$
$$+ \rho\omega^2 R_P^2 \{ -\tfrac{1}{2}(R_B/r)^2 + \tfrac{1}{2}(r/R_B)^2 - 2\log(r/R_B) + constant\}/$$
$$(R_B/R_P - R_P/R_B)^2 \qquad (3-38)$$

Observe that the pressure $p(r,\theta,z)$ depends on all three coordinates, even though $v_z(r)$ depends only on r. The pressure gradient $\partial p/\partial r$, for example, throws cuttings through centrifugal force; it likewise depends on r, θ and z, and also on ρ, g and α. It may be an important correlation parameter in cuttings transport and bed formation studies. The additive constants in Equations 3-35 and 3-38 have no dynamical significance. Equations 3-33 to 3-38 describe completely and *exactly* the internal drillpipe flow and the external annular borehole flow. No geometrical simplifications have been made. The solution applies to an inclined, centered drillstring rotating at a constant angular rate ω, but it is restricted to a Newtonian fluid. Again, these concentric solutions show that in the Newtonian limit, the velocities $v_z(r)$ and $v_\theta(r)$ uncouple; this is not the case for eccentric flows. And this is never so with non-Newtonian drilling flows, concentric or eccentric. Thus, the analysis methods developed here must be extended to account for the physical coupling.

NARROW ANNULUS POWER LAW SOLUTION

For general non-Newtonian flows, the Navier-Stokes equations (see Equations 3-17 to 3-19) do not apply; direct recourse to Equations 3-1 to 3-3 must be made. However, many of the physical assumptions used and justified above still hold. If we again assume a constant density flow, and also that velocities do not vary with z, θ and t, and that $v_r = 0$, we again obtain our Equation 3-24. This implies mass conservation. It leads to further simplifications in Equations 3-1 to 3-3, and in the tensor definitions given by Equations 3-5 to 3-14. The result is the reduced system of equations

$$0 = \rho g \sin \alpha - \partial p/\partial z + 1/r\, \partial(Nr\, \partial v_z/\partial r)/\,\partial r \qquad (3-39)$$

$$0 = -\rho g \cos \alpha \cos \theta - 1/r\, \partial p/\partial\theta + 1/r^2\, \partial(Nr^3\, \partial(v_\theta/r)\, \partial r)/\partial r \qquad (3-40)$$

$$-\rho v_\theta^2/r = -\rho g \cos \alpha \sin \theta - \partial p/\partial r$$
$$+ 1/r\, \partial(Nr\, \partial(v_\theta/r)/\,\partial r)/\,\partial\theta + \partial(N\, \partial v_z/\partial r)/\partial z \qquad (3-41)$$

At this point, we introduce the same separation of variables for pressure used for Newtonian flows, that is Equation 3-29, so that Equations 3-39 to 3-41 become

$$0 = -\partial P/\partial z + 1/r\, \partial\, (Nr\, \partial v_z/\partial r)/\, \partial r \qquad\qquad (3\text{-}42)$$

$$0 = \partial\, (Nr^3\, \partial\, (v_\theta/r)/\, \partial r)/\, \partial r \qquad\qquad (3\text{-}43)$$

$$-\rho v_\theta^2/r = -\partial P^*/\partial r + 1/r\, \partial\, (Nr\, \partial\, (v_\theta/r)/\, \partial r)/\, \partial\theta$$
$$+\, \partial\, (N\, \partial v_z/\partial r)/\, \partial z \qquad\qquad (3\text{-}44)$$

Of course, the $P^*(r)$ applicable to non-Newtonian flows will follow from the solution to Equation 3-44; Equations 3-35 and 3-38 for Newtonian flows do not apply. Since θ does not explicitly appear in Equation 3-44, v_z and v_θ do not depend on θ; and on z either, as previously assumed. Thus, all partial derivatives with respect to θ and z vanish. Without approximation, the final set of *ordinary* differential equations takes the form

$$1/r\, d(Nr\, dv_z/dr)/dr = dP/dz = \text{constant} \qquad\qquad (3\text{-}45)$$

$$d(Nr^3\, d(v_\theta/r)dr)/dr = 0 \qquad\qquad (3\text{-}46)$$

$$dP^*/dr = \rho v_\theta^2/r \qquad\qquad (3\text{-}47)$$

where $N(\Gamma)$ is the complete velocity functional given in Equation 3-16. The application of Equation 3-16 couples our axial and azimuthal velocities, and is the source of mathematical complication.

The solution to Equations 3-45 to 3-47 may appear to be simple. For example, the unknowns v_θ and v_z are governed by two second-order ordinary differential equations, namely, Equations 3-45 and 3-46; the four constants of integration are completely determined by four no-slip conditions at the rotating drillstring surface and the stationary borehole wall. And, the radial pressure (governed by Equation 3-47) is obtained after the fact only, once v_θ is available.

In reality, the difficulty lies with the fact that Equations 3-45 and 3-46 are nonlinearly coupled through Equation 3-16. It is not possible to solve for either v_z and v_θ sequentially, as we did for the "simpler" Navier-Stokes equations. Because the actual physical coupling is strong at the leading order, it is incorrect to solve for non-Newtonian effects using perturbation series methods, say, expanded about decoupled Newtonian solutions. The method described here required tedious trial and error; 24 ways to implement no-slip conditions were possible, and not all yielded integrable equations.

We successfully derived closed form, explicit, analytical solutions for the coupled velocity fields. However, the desire for closed form solutions required an additional "narrow annulus" assumption. Still, the resulting solutions are useful since they yield explicit answers for rotating flows, thus providing key physical insight into the role of different flow parameters.

The method devised for arbitrary n below does *not* apply to the Newtonian limit where n = 1, for which solutions are already available. But in the n → 1 ± limit of our power law results, we will show that we recover the Navier-Stokes solution. Thus, the physical dependence on n is continuous, and the results obtained in this chapter cover all values of n. With these preliminary remarks said and done, we proceed with the analysis.

Let us multiply Equation 3-45 by r throughout. Next, integrate the result, and also integrate Equation 3-46 once with respect to r, to yield

$$Nr \, dv_z/dr = r^2/2 \, dP/dz + E_1 \qquad (3\text{-}48)$$

$$Nr^3 \, d(v_\theta/r)dr = E_2 \qquad (3\text{-}49)$$

where E_1 and E_2 are integration constants. Division of Equation 3-48 by Equation 3-49 gives a result (independent of the apparent viscosity $N(\Gamma)$) relating v_z to v_θ/r, namely,

$$dv_z/dr = (r^4/2 \, dP/dz + E_1 r^2)/E_2 \, d(v_\theta/r)dr \qquad (3\text{-}50)$$

At this point, it is convenient to introduce the angular velocity

$$\Omega(r) = v_\theta/r \qquad (3\text{-}51)$$

Substitution of the tensor elements $\underline{D}$ in Equation 3-13 leads to

$$\Gamma = \{ 2 \text{ trace } (\underline{D} \bullet \underline{D}) \}^{1/2}$$
$$= [(\partial v_z/\partial r)^2 + r^2 \, (\partial \{v_\theta/r\}/\partial r)^2]^{1/2} \qquad (3\text{-}52)$$

so that the power law apparent viscosity given by Equation 3-14 becomes

$$N(\Gamma) = k \, [(\partial v_z/\partial r)^2 + r^2 \, (\partial \{v_\theta/r\}/\partial r)^2]^{(n-1)/2} \qquad (3\text{-}53)$$

These results were stated without proof in Equations 3-15 and 3-16. Now combine Equations 3-49 and 3-53 so that

$$k \, [(\partial v_z/\partial r)^2 + r^2 \, (\partial \Omega/\partial r)^2]^{(n-1)/2} \, d\Omega/dr = E_2/r^3 \qquad (3\text{-}54)$$

If dv_z/dr is eliminated using Equation 3-50, we obtain, after very lengthy manipulations,

$$d\Omega/dr = (E_2/k)^{1/n} \, [r^{(2n+4)/(n-1)}$$
$$+ r^{(4n+2)/(n-1)} \, \{(E_1 + r^2/2 \, dP/dz)/E_2\}^2]^{(1-n)/2n} \qquad (3\text{-}55)$$

Next, integrate Equation 3-55 over the interval (r, R_B) where R_B is the borehole radius. If we apply the first no-slip boundary condition

$$\Omega(R_B) = 0 \qquad (3\text{-}56)$$

(there are four no-slip conditions altogether) and invoke the Mean Value Theorem of differential calculus, using as the appropriate mean the arithmetic average, we obtain

$$\Omega\,(r) = (E_2/k)^{1/n}\,(r-R_B)\,[((r+R_B)/2)^{(2n+4)/(n-1)}$$

$$+((r+R_B)/2)^{(4n+2)/(n-1)}\,\{(E_1 + (r+R_B)^2/8\ dP/dz)/E_2\}^2\,]^{(1-n)/2n} \qquad (3\text{-}57)$$

At this point, though, we do not yet apply any of the remaining three no-slip velocity boundary conditions.

We turn our attention to v_z instead. We can derive a differential equation independent of Ω by combining Equations 3-50, 3-51 and 3-55 as follows,

$$dv_z/dr = (r^4/2\ dP/dz + E_1r^2)/E_2\ d\Omega/dr \qquad (3\text{-}58)$$

$$= r^2(E_1 + r^2/2\ dP/dz)/E_2 \times (E_2/k)^{1/n}\,[r^{(2n+4)/(n-1)}$$

$$+ r^{(4n+2)/(n-1)}\,\{(E_1 + r^2/2\ dP/dz)/E_2\}^2\,]^{(1-n)/2n}$$

We next integrate Equation 3-58 over (R_P, r), where R_P is the drillpipe radius, subject to the second no-slip condition

$$v_z\,(R_P) = 0 \qquad (3\text{-}59)$$

An integration similar to that used for Equation 3-55, again invoking the Mean Value Theorem, leads to a result analogous to Equation 3-57, that is,

$$v_z\,(r) = ((r+R_P)/2)^2(E_1 + ((r+R_P)/2)^2/2\ dP/dz)/E_2 \times$$

$$(E_2/k)^{1/n}\,[((r+R_P)/2)^{(2n+4)/(n-1)} + ((r+R_P)/2)^{(4n+2)/(n-1)}$$

$$\{(E_1 + ((r+R_P)/2)^2/2\ dP/dz)/E_2\}^2\,]^{(1-n)/2n}\,(r - R_P) \qquad (3\text{-}60)$$

Very useful results are obtained if we now apply the third no-slip condition

$$v_z\,(R_B) = 0 \qquad (3\text{-}61)$$

With this constraint, Equation 3-60 leads to a somewhat unwieldy combination of terms, namely,

$$0 = ((R_B+R_P)/2)^2(E_1 + ((R_B+R_P)/2)^2/2\ dP/dz)/E_2 \times$$

$$(E_2/k)^{1/n}\,[((R_B+R_P)/2)^{(2n+4)/(n-1)} + ((R_B+R_P)/2)^{(4n+2)/(n-1)}$$

$$\{(E_1 + ((R_B+R_P)/2)^2/2\ dP/dz)/E_2\}^2\,]^{(1-n)/2n}\,(R_B - R_P) \qquad (3\text{-}62)$$

But if we observe that the quantity contained within the square brackets "[ ]" is positive definite, and that $(R_B - R_P)$ is nonzero, it follows that the left-hand side "0" can be obtained only if

$$E_1 = -(R_B + R_P)^2/8 \; dP/dz \tag{3-63}$$

holds identically. The remaining integration constant E_2 is determined from the last of the four no-slip conditions

$$\Omega(R_P) = \omega \tag{3-64}$$

Equation 3-64 requires fluid at the pipe surface to move with the rotating surface. Here, without loss of generality, $\omega < 0$ is the constant drillstring angular rotation speed. Combination of Equations 3-57, 3-63 and 3-64, after lengthy manipulations, leads to the surprisingly simple result that

$$E_2 = k \; (\omega/(R_P - R_B))^n \; ((R_P + R_B)/2)^{n+2} \tag{3-65}$$

With all four no-slip conditions applied, the four integration constants, and hence the analytical solution for our power law model, are completely determined. We next perform validation checks before deriving applications formulas.

ANALYTICAL VALIDATION

Different analytical procedures were required for Newtonian flows and power law flows with general n's. This is related to the decoupling between axial and circumferential velocities in the singular $n = 1$ limit. On physical grounds, we expect that the power law solution, if correct, would behave "continuously" through $n = 1$ as the fluid passes from dilatant to pseudoplastic states. That is, the solution should change smoothly when n varies from $1-\delta$ to $1+\delta$ where $|\delta| \ll 1$ is a small number. This continuous dependence and physical consistency will be demonstrated next. This validation also guards against error, given the quantity of algebraic manipulations involved.

The formulas derived above for general power law fluids will be checked against exact Newtonian results where $k = \mu$ and $n = 1$. For consistency, we will take the narrow annulus limit of those formulas, a geometric approximation used in the power law derivation. We will demonstrate that the closed form results obtained for non-Newtonian fluids are indeed "continuous in n" through the singular point $n = 1$.

We first check our results for the stresses $S_{r\theta}$ and $S_{\theta r}$. From Equations 3-5, 3-9 and 3-57, we find that

$$S_{r\theta} = S_{\theta r} = k \; (\omega/(R_P - R_B))^n \; ((R_P + R_B)/2)^{n+2} \; r^{-2} \tag{3-66}$$

In the limit $k = \mu$ and $n = 1$, Equation 3-66 for power law fluids reduces to

$$S_{r\theta} = S_{\theta r} = \mu \; \omega/\{(R_P - R_B)r^2\} \times ((R_P + R_B)/2)^3 \tag{3-67}$$

On the other hand, the definition $S_{r\theta} = S_{\theta r} = \mu \, d\Omega/dr$ inferred from Equations 3-5 and 3-9 becomes, using Equations 3-37 and 3-51 for Newtonian flow,

$$S_{r\theta} = S_{\theta r} = \mu \, \omega/\{(R_P - R_B)r^2\} \times 2(R_P R_B)^2/(R_P + R_B) \quad (3\text{-}68)$$

Are the two second factors "$((R_P+R_B)/2)^3$" and "$2(R_P R_B)^2/(R_P +R_B)$" in Equations 3-67 and 3-68 consistent? If we evaluate these expressions in the narrow annulus limit, setting $R_P = R_B = R$, we obtain R^3 in *both* cases, providing the required validation. This consistency holds for all values of dP/dz.

For our second check, consider the power law stresses S_{rz} and S_{zr} obtained from Equations 3-5, 3-10 and 3-58, that is,

$$S_{rz} = S_{zr} = E_1/r + \tfrac{1}{2} r \, dP/dz$$

$$= \{\tfrac{1}{2} r - (R_P+R_B)^2/(8r)\} \, dP/dz \quad (3\text{-}69)$$

The corresponding formula in the Newtonian limit is

$$S_{rz} = S_{zr} = \mu \, dv_z/dr$$

$$= \{\tfrac{1}{2} r - (R_P^2 - R_B^2)/(4r \log R_P/R_B)\} \, dP/dz \quad (3\text{-}70)$$

where we have used Equation 3-36. Now, is "$(R_P+R_B)^2/8$" consistent with "$(R_P^2 -R_B^2)/(4 \log R_P/R_B)$"? As before, consider the narrow annulus limit, setting $R_P = R_B = R$. The first expression easily reduces to $R^2/2$. For the second, we expand $\log R_B/R_P = \log \{1 + (R_B-R_P)/R_P\} = (R_B-R_P)/R_P$ and retain only the first term of the Taylor expansion. Direct substitution yields $R^2/2$ again. Therefore, Equations 3-69 and 3-70 are consistent for all rotation rates ω. Thus, from our checks on both S_{rz} and $S_{r\theta}$, we find good physical consistency and consequently reliable algebraic computations.

DIFFERENCES BETWEEN NEWTONIAN AND POWER LAW FLOWS

Equations 3-29, 3-51, 3-57, 3-60, 3-63 and 3-65 specify the velocity fields v_z and $v_\theta = r\Omega(r)$ as functions of wellbore geometry, fluid rheology, pipe inclination, rotation rate, pressure gradient and gravity. We emphasize that Equation 3-47, which is to be evaluated using the non-Newtonian solution for v_θ, provides only a partial solution for the complete radial pressure gradient. The remaining part is obtained by adding the "$- \rho g \cos \alpha \sin \theta$" contribution of Equation 3-29. As in Newtonian flows, the pressure and its spatial gradients depend on all the coordinates r, θ and z, and the parameters ρ, g and α.

There are fundamental differences between these solutions and the Newtonian ones. For example, in the latter, the solutions for v_z and v_θ completely decouple despite the nonlinearity of the Navier-Stokes equations. The governing equations become linear. But for power law flows, both v_z and v_θ remain highly coupled and nonlinear. In this sense, Newtonian results are singular; but the degeneracy disappears for eccentric geometries when the convective terms reappear. Cuttings transport experimenters working with *concentric Newtonian* flows will *not* be able to extrapolate their findings to practical geometries or fluids. This will be discussed in detail in Chapter 5.

Also, the expression for v_z in the Newtonian limit is directly proportional to dP/dz; but as Equation 3-60 for power law fluids shows, the dependence of v_z (and hence, of total volume flow rate) on pressure gradient is a nonlinear one. Similarly, while Equation 3-37 shows that v_θ is directly proportional to the rotation rate ω, Equations 3-51, 3-57 and 3-65 illustrate a more complicated nonlinear dependence for power law fluids. It is important to emphasize that, for a fixed annular flow geometry in Newtonian flow, v_z depends only on dP/dz and not ω, and v_θ depends only on ω and not dP/dz. But for power law flows, v_z and v_θ each depend on both dP/dz and ω. Thus, "axial quantities" like net annular volume flow rate cannot be calculated without considering both dP/dz and ω.

Interestingly, though, the stresses $S_{r\theta}$ and S_{rz} in the non-Newtonian case preserve their "independence" as in Newtonian flows. That is, $S_{r\theta}$ depends only on ω and not dP/dz, while S_{rz} depends only on dP/dz and not ω (see Equations 3-71 to 3-74 below). The power law stress values themselves, of course, are different from the Newtonian counterparts. And also, the "maximum stress" $(S_{r\theta}^2 + S_{rz}^2)^{1/2}$, important in borehole stability and cuttings bed erosion, depends on both ω and dP/dz, as it does in Newtonian flow.

An important question is the significance of rotation in practical calculations. Can "ω" be safely neglected in drilling and cementing applications? This depends on a nondimensional ratio of circumferential to axial momentum flux. While the "maximum v_θ" is easily obtained as "$\omega_{rpm} \times R_p$," the same estimate for v_z is difficult to obtain since axial velocity is sensitive to both n and k, not to mention v_θ and dP/dz. In general, one needs to consider the full problem without approximation.

Of course, since the analytical solution is now available, the use of approximate "axial flow only" solutions is really a moot point. The power law results and the formulas derived next are "explicit" in that they require no iteration. And although the software described later is written in Fortran, our equations are just as easily programmed on calculators. The important dependence of annular flows on "ω" will be demonstrated in calculated results.

MORE APPLICATIONS FORMULAS

The cylindrical geometry of the present problem renders all stress tensor components except $S_{r\theta}$, $S_{\theta r}$, S_{zr} and S_{rz} zero. From our power law results, the required formulas for viscous stress can be shown to be

$$S_{r\theta} = S_{\theta r} = k \, (\omega/(R_P - R_B))^n \, ((R_P+R_B)/2)^{n+2} \, r^{-2} \qquad (3\text{-}71)$$

$$S_{rz} = S_{zr} = E_1/r + \tfrac{1}{2} r \, dP/dz$$

$$= \{\tfrac{1}{2} r - (R_P+R_B)^2/(8r)\} \, dP/dz \qquad (3\text{-}72)$$

Their Newtonian counterparts take the form

$$S_{r\theta} = S_{\theta r} = \mu \, \omega/\{(R_P - R_B)r^2\} \times 2(R_P R_B)^2/(R_P + R_B) \qquad (3\text{-}73)$$

$$S_{rz} = S_{zr} = \mu \, dv_z/dr$$

$$= \{\tfrac{1}{2} r - (R_P^2 - R_B^2)/(4r \log R_P/R_B)\} \, dP/dz \qquad (3\text{-}74)$$

In studies on borehole erosion, annular velocity plays an important role, since drilling mud carries abrasive cuttings. The magnitude of fluid shear stress may also be important in unconsolidated sands where tangential surface forces assist in wall erosion. Stress considerations also arise in cuttings bed transport analysis in highly deviated or horizontal holes (see Chapter 5). The individual components can be obtained by evaluating Equations 3-71 and 3-72 at $r = R_B$ for power law fluids, and Equations 3-73 and 3-74 for Newtonian fluids. And since these stresses act in orthogonal directions, the "maximum stress" can be obtained by writing

$$S_{max}(R_B) = \{S_{r\theta}^2(R_B) + S_{rz}^2(R_B)\}^{1/2} \qquad (3\text{-}75)$$

The shear force associated with this stress acts in a direction offset from the borehole axis by an angle

$$\Theta_{\text{max shear}} = \arctan \{S_{r\theta}(R_B)/S_{rz}(R_B)\} \qquad (3\text{-}76)$$

Opposing the erosive effects of shear may be the stabilizing effects of hydrostatic and dynamic pressure. Explicit formulas for the pressures $P(z)$, $P^*(r)$ and the hydrostatic background level were given earlier.

To obtain the corresponding elements of the deformation tensor, we rewrite Equation 3-5 in the form

$$\underline{D} = \underline{S} / 2N(\Gamma) \qquad (3\text{-}77)$$

and substitute S_{rz} or $S_{r\theta}$ as required. In the Newtonian case, $N(\Gamma) = \mu$ is the laminar viscosity; for power law fluids, Equation 3-16 applies. Stresses are important to transport problems; fluid deformations are useful for the kinematic studies often of interest to rheologists.

Annular volume flow rate Q as it depends on pressure gradient is important in determining mud pump power requirements and cuttings transport capabilities of the drilling fluid. It is obtained by evaluating

$$Q = \int_{R_P}^{R_B} v_z(r)\, 2\pi r\, dr \qquad (3\text{-}78)$$

In the above integrand, Equation 3-36 for v_z (r) must be used for Newtonian flows, while Equation 3-60 would apply to power law fluids.

Borehole temperature may play an important role in drilling. Problem areas include formation temperature interpretation and mud thermal stability (e.g., the "thinning" of oil-base muds with temperature limits cuttings transport efficiency). Many studies do not consider the effects of heat generation by internal friction, which may be non-negligible; in closed systems, temperature increases over time may be significant. Ideally, temperature effects due to fluid type and cumulative effects related to total circulation time should be identified.

The starting point is the equation describing energy balance within the fluid, that is, the PDE for the temperature field $T(r,\theta,z,t)$. Even if the velocity field is steady, temperature effects will typically not be, since irreversible thermodynamic effects cause continual increases of T with time. If temperature increases are large enough, the changes of viscosity, consistency factor or fluid exponent as functions of T must be considered. Then the momentum and energy equations will be coupled. We will not consider this complicated situation yet, so that the velocity fields can be obtained independently of T. For Newtonian flows, we have $n = 1$ and $k = \mu$. The temperature field satisfies

$$\rho c\, (\partial T/\partial t + v_r\, \partial T/\partial r + v_\theta/r\, \partial T/\partial \theta + v_z\, \partial T/\partial z) = \qquad (3\text{-}79)$$

$$= K\, [\, 1/r\, \partial\, (r\, \partial T/\partial r)/\, \partial r + 1/r^2\, \partial^2 T/\partial \theta^2 + \partial^2 T/\partial z^2\,]$$

$$+ 2\mu\, \{\, (\partial v_r/\partial r)^2 + [1/r\, (\partial v_\theta/\partial \theta + v_r)]^2 + (\partial v_z/\partial z)^2\, \}$$

$$+ \mu\, \{\, (\partial v_\theta/\partial z + 1/r\, \partial v_z/\partial \theta)^2 + (\partial v_z/\partial r + \partial v_r/\partial z)^2$$

$$+ [1/r\, \partial v_r/\partial \theta + r\, \partial\, (v_\theta/r)/\, \partial r]^2\, \}$$

$$+ \rho Q^*$$

where c is the heat capacity, K is the thermal conductivity, and Q^* is an energy transmission function. The terms on the first line represent transient and convective effects; the second line models heat conduction. Those on the third through fifth are positive definite and represent the heat generation due to internal fluid friction. These irreversible thermodynamic effects are referred to collectively as the "dissipation function" or "heat generation function." The dissipation function Φ is in effect a distributed heat source within the moving fluid medium. If we employ the same assumptions as used in our solution of the Navier-Stokes equations for Newtonian flows, this expression reduces to

$$\Phi = \mu \left\{ (\partial v_z/\partial r)^2 + r^2(\partial \Omega/\partial r)^2 \right\} > 0 \qquad (3\text{-}80)$$

which can be easily evaluated using Equations 3-36, 3-37 and 3-51. It is important to recognize that Φ depends on spatial velocity gradients only, and not on velocity magnitudes. In a closed system, the fact that $\Phi > 0$ leads to increases of temperature in time if the borehole walls cannot conduct heat away quickly. Equations 3-79 and 3-80 assume Newtonian flow. For general fluids, it is possible to show that the dissipation function now takes the specific form

$$\begin{aligned}
\Phi = {} & S_{rr}\, \partial v_r/\partial r + S_{\theta\theta}\, 1/r\,(\partial v_\theta/\partial\theta + v_r) \\
& + S_{zz}\, \partial v_z/\partial z + S_{r\theta}\, [r\,\partial\,(v_\theta/r)/\,\partial r + 1/r\,\partial v_r/\partial\theta] \\
& + S_{rz}\,(\partial v_z/\partial r + \partial v_r/\partial z) + S_{\theta z}\,(1/r\,\partial v_z/\partial\theta + \partial v_\theta/\partial z) \qquad (3\text{-}81)
\end{aligned}$$

The geometrical simplifications used earlier reduce Equation 3-81 to

$$\Phi = k \left\{ (\partial v_z/\partial r)^2 + r^2(\partial\Omega/\partial r)^2 \right\}^{(n+1)/2} > 0 \qquad (3\text{-}82)$$

In the Newtonian limit with $k = \mu$ and $n = 1$, Equation 3-82 consistently reduces to Equation 3-80. Equations 3-55 and 3-58 are used to evaluate the expression for Φ above. As before, Φ depends upon velocity gradients only and not magnitudes; it largely arises from high shear at solid boundaries.

DETAILED CALCULATED RESULTS

The power law results derived above were coded in a Fortran algorithm designed to provide a suite of output "utility" solutions for any set of input data. These may be useful in determining operationally important quantities like volume flow rate and axial speed. But they also provide research utilities needed, for example, to correlate experimental cuttings transport data or interpret formation temperature data.

The core code resides in 30 lines of Fortran. It runs on a "stand alone" basis or as an embedded subroutine for specialized applications. The formulas used are also programmable on calculators. Inputs include pipe or casing outer diameter, borehole diameter, axial pressure gradient, rotation rate, fluid exponent n, and consistency factor k. Outputs include tables, line plots, and ASCII character plots versus "r" for a number of useful functions. These are,

- o   Axial velocity $v_z\,(r)$
- o   Circumferential velocity $v_\theta\,(r)$
- o   Fluid rotation rate $\omega(r)$ ("local rpm")

o   Total absolute speed
o   Angle between v_z (r) and v_θ (r)
o   Axial velocity gradient dv_z (r)/dr
o   Azimuthal velocity gradient dv_θ (r)/dr
o   Angular velocity gradient $d\omega(r)/dr$
o   Radial pressure gradient
o   Apparent viscosity versus "r"
o   Local frictional heat generation
o   All stress tensor components
o   Maximum wellbore stress
o   All deformation tensor components

We emphasize that the "radial pressure gradient" above refers to the partial contribution in Equation 3-47, which depends on "r" only. For the complete gradient, Equation 3-29 shows that the term "- $\rho g \cos \alpha \sin \theta$" must be appended to the value calculated here. This contribution depends on ρ, g, α and θ. In addition to the foregoing arrays, total annular volume flow rate and radial averages of all of the above quantities are computed. Before proceeding to detailed computations, let us compare our concentric, rotating pipe, *narrow annulus* results in the limit of zero rotation with an exact solution.

Example 1. East Greenbriar No. 2

A mud hydraulics analysis was performed for "East Greenbriar No. 2" using a computer program offered by a service company. This program, which applies to nonrotating flows only, is based on the *exact* Fredrickson and Bird (1958) solution. In this example, the drillpipe outer radius is 2.5 in, the borehole radius is 5.0 in, the axial pressure gradient is 0.00389 psi/ft, the fluid exponent is 0.724, and the consistency factor is 0.268 lbf sec$^{0.724}$ / (100 ft^2) (that is, 0.1861 × 10^{-4} lbf sec$^{0.724}$ /in^2 in the units employed by our program). The exact results computed using this data are an annular volume flow rate of 400 gal/min, and an average axial speed of 130.7 ft/min. The same input data was used in our program, with an assumed drillstring "rpm" of 0.001. We computed 373.6 gal/min and 126.9 ft/min for this nonrotating flow, agreeing to within 7% for the not-so-narrow annulus.

Our model was designed, of course, to include the effects of drillstring rotation. We first considered an extremely large rpm of 300, with the same pressure gradient, to evaluate qualitative effects. The corresponding results were 526.1 gal/min and 175.6 ft/min. The ratio of the average circumferential speed to the average axial speed is 1.06, indicating that rotational effects are important. At 150 rpm, our volume flow rate of 458.7 gal/min exceeds 373.6 gal/min by 23%. In this case, the ratio of average circumferential speed to axial speed is still a non-negligible 65%. These results suggest that static models tend to overestimate the pressure requirements needed by a rotating drillstring to produce a prescribed flow rate. Our hydraulics model indicates that including rotational effects, for a fixed pressure gradient, is likely to increase the volume flow rate over static predictions. These considerations may be important in planning long deviated wells where one needs to know, for a given rpm, what maximum borehole length is possible with the pump at hand.

Example 2. Detailed Spatial Properties Versus "r"

Our computational algorithm does more than calculate annular volume flow rate and average axial speed. This section includes the entire output file from a typical run, in this case "East Greenbriar No. 2," with annotated comments. The input menu is nearly identical to the summary in Table 3-1. Because the numerical results are based on analytical, closed form results, there are no computational inputs; the grid reference in Table 3-1 is a print control parameter. At the present, the volume flow rate is the only quantity computed numerically; a second-order scheme is applied to our $v_z(r)$'s. All inputs are in "plain English" and are easily understandable. Outputs are similarly "user friendly." All output quantities are defined, along with units, in a printout that precedes tabulated and plotted results. This printout is duplicated in Table 3-2.

Table 3-1
Summary of Input Parameters

```
O  Drill pipe outer radius (inches) = 2.5000
O  Borehole radius (inches) = 5.0000
O  Axial pressure gradient (psi/ft) = 0.0039
O  Drillstring rotation rate (rpm) = 300.0000
O  Drillstring rotation rate (rad/sec) = 31.4159
O  Fluid exponent "n" (nondimensional) = 0.7240
O  Consistency factor (lbf sec^n/sq in) = 0.1861E-04
O  Mass density of fluid (lbf^2 sec^4/ft ) = 1.9000
   (e.g., about 1.9 for water)
O  Number of radial "grid" positions = 18
```

Table 3-2
Analytical (Non-Iterative) Solutions
Tabulated versus "r," Nomenclature and Units

r	Annular radial position	(in)
V_z	Velocity in axial z direction	(in/sec)
V_θ	Circumferential velocity	(in/sec)
$d\theta/dt$ or W	θ velocity	(rad/sec)
	(Note: 1 rad/sec = 9.5493 rpm)	
dV_z/dr	Velocity gradient	(1/sec)
dV_θ/dr	Velocity gradient	(1/sec)
dW/dr	Angular speed gradient	$(1/(\text{sec} \times \text{in}))$
$S_{r\theta}$	rθ stress component	(psi)
S_{rz}	rz stress component	(psi)
S_{max}	Sqrt $(S_{rz}{**}2 + S_{r\theta}{**}2)$	(psi)
dP/dr	Radial pressure gradient	(psi/in)
App-Vis	Apparent viscosity	(lbf sec /sq in)
Dissip	Dissipation function	$(\text{lbf}/(\text{sec} \times \text{sq in}))$
	(indicates frictional heat produced)	
Atan V_θ/V_z	Angle between V_θ and V_z vectors	(deg)
Net Spd	Sqrt $(V_z{**}2 + V_\theta{**}2)$	(in/sec)
$D_{r\theta}$	rθ deformation tensor component	(1/sec)
D_{rz}	rz deformation tensor component	(1/sec)

Table 3-3
Calculated Quantities vs "r"

r	V_z	V_θ	W	$d(V_z)/dr$	$d(V_\theta)/dr$	dW/dr
5.00	.601E-04	.279E-04	.559E-05	-.610E+02	-.293E+02	-.586E+01
4.86	.848E+01	.407E+01	.837E+00	-.534E+02	-.293E+02	-.620E+01
4.72	.164E+02	.814E+01	.172E+01	-.460E+02	-.294E+02	-.659E+01
4.58	.237E+02	.122E+02	.266E+01	-.390E+02	-.295E+02	-.702E+01
4.44	.304E+02	.163E+02	.366E+01	-.321E+02	-.297E+02	-.751E+01
4.31	.365E+02	.203E+02	.472E+01	-.256E+02	-.302E+02	-.811E+01
4.17	.418E+02	.244E+02	.585E+01	-.193E+02	-.310E+02	-.885E+01
4.03	.462E+02	.284E+02	.705E+01	-.131E+02	-.324E+02	-.980E+01
3.89	.497E+02	.325E+02	.835E+01	-.672E+01	-.345E+02	-.110E+02
3.75	.521E+02	.366E+02	.975E+01	.273E-04	-.374E+02	-.126E+02
3.61	.533E+02	.407E+02	.113E+02	.738E+01	-.412E+02	-.145E+02
3.47	.532E+02	.449E+02	.129E+02	.157E+02	-.461E+02	-.170E+02
3.33	.516E+02	.492E+02	.148E+02	.251E+02	-.521E+02	-.201E+02
3.19	.483E+02	.536E+02	.168E+02	.358E+02	-.594E+02	-.239E+02
3.06	.432E+02	.582E+02	.190E+02	.480E+02	-.682E+02	-.286E+02
2.92	.361E+02	.630E+02	.216E+02	.619E+02	-.787E+02	-.344E+02
2.78	.266E+02	.680E+02	.245E+02	.778E+02	-.914E+02	-.417E+02
2.64	.147E+02	.732E+02	.277E+02	.959E+02	-.107E+03	-.510E+02
2.50	.000E+00	.785E+02	.314E+02	.117E+03	-.126E+03	-.628E+02

Table 3-3
Calculated Quantities vs "r" (continued)

r	$s_{r\theta}$	s_{rz}	s_{max}	dP/dr	App-Vis	Dissip
5.00	.170E-03	-.355E-03	.393E-03	.143E-13	.582E-05	.266E-01
4.86	.180E-03	-.319E-03	.366E-03	.312E-03	.598E-05	.225E-01
4.72	.191E-03	-.283E-03	.341E-03	.128E-02	.614E-05	.190E-01
4.58	.203E-03	-.246E-03	.318E-03	.298E-02	.630E-05	.161E-01
4.44	.216E-03	-.208E-03	.299E-03	.545E-02	.646E-05	.139E-01
4.31	.230E-03	-.168E-03	.285E-03	.878E-02	.658E-05	.123E-01
4.17	.245E-03	-.128E-03	.277E-03	.131E-01	.665E-05	.115E-01
4.03	.262E-03	-.869E-04	.277E-03	.184E-01	.665E-05	.115E-01
3.89	.282E-03	-.442E-04	.285E-03	.248E-01	.658E-05	.124E-01
3.75	.303E-03	.175E-09	.303E-03	.326E-01	.643E-05	.143E-01
3.61	.327E-03	.459E-04	.330E-03	.420E-01	.622E-05	.175E-01
3.47	.353E-03	.936E-04	.365E-03	.532E-01	.598E-05	.223E-01
3.33	.383E-03	.144E-03	.409E-03	.665E-01	.573E-05	.292E-01
3.19	.417E-03	.196E-03	.461E-03	.824E-01	.547E-05	.388E-01
3.06	.456E-03	.251E-03	.520E-03	.102E+00	.523E-05	.518E-01
2.92	.501E-03	.309E-03	.588E-03	.125E+00	.499E-05	.693E-01
2.78	.552E-03	.370E-03	.665E-03	.152E+00	.476E-05	.928E-01
2.64	.611E-03	.436E-03	.751E-03	.186E+00	.455E-05	.124E+00
2.50	.681E-03	.507E-03	.849E-03	.226E+00	.434E-05	.166E+00

Table 3-3
Calculated Quantities vs "r" (continued)

r	V_z	V_θ	Atan V_θ/V_z	Net Spd	$D_{r\theta}$	D_{rz}
5.00	.601E-04	.279E-04	.249E+02	.663E-04	.146E+02	-.305E+02
4.86	.848E+01	.407E+01	.256E+02	.940E+01	.151E+02	-.267E+02
4.72	.164E+02	.814E+01	.264E+02	.183E+02	.155E+02	-.230E+02
4.58	.237E+02	.122E+02	.272E+02	.267E+02	.161E+02	-.195E+02
4.44	.304E+02	.163E+02	.281E+02	.345E+02	.167E+02	-.161E+02
4.31	.365E+02	.203E+02	.291E+02	.417E+02	.175E+02	-.128E+02
4.17	.418E+02	.244E+02	.303E+02	.483E+02	.184E+02	-.965E+01
4.03	.462E+02	.284E+02	.316E+02	.542E+02	.197E+02	-.653E+01
3.89	.497E+02	.325E+02	.332E+02	.593E+02	.214E+02	-.336E+01
3.75	.521E+02	.366E+02	.351E+02	.636E+02	.236E+02	.137E-04
3.61	.533E+02	.407E+02	.374E+02	.670E+02	.262E+02	.369E+01
3.47	.532E+02	.449E+02	.402E+02	.696E+02	.295E+02	.783E+01
3.33	.516E+02	.492E+02	.436E+02	.713E+02	.334E+02	.125E+02
3.19	.483E+02	.536E+02	.480E+02	.722E+02	.381E+02	.179E+02
3.06	.432E+02	.582E+02	.534E+02	.725E+02	.436E+02	.240E+02
2.92	.361E+02	.630E+02	.602E+02	.726E+02	.502E+02	.309E+02
2.78	.266E+02	.680E+02	.686E+02	.730E+02	.579E+02	.389E+02
2.64	.147E+02	.732E+02	.786E+02	.746E+02	.673E+02	.480E+02
2.50	.000E+00	.785E+02	.900E+02	.785E+02	.785E+02	.584E+02

The defined quantities are first tabulated, as shown in Table 3-3, as a function of the radial position "r." At this point, the total volume flow rate is computed and presented in textual form, that is,

```
Total volume flow rate (cubic in/sec) =  .2026E+04
                       (gal/min) =  .5261E+03
```

A run-time screen menu prompts the user with regard to quantities he would like displayed in ASCII file plots. The complete list of quantities was given previously. Plots corresponding to "East Greenbriar No. 2" are shown next with annotations.

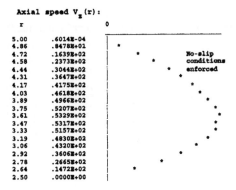

Figure 3-2. Axial speed.

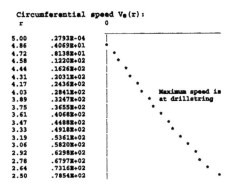

Figure 3-3. Circumferential speed.

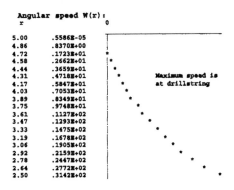

Figure 3-4. Angular speed.

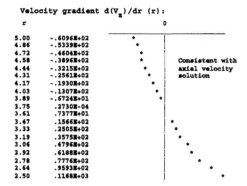

Velocity gradient d(V$_z$)/dr (r):

r		0
5.00	-.6096E+02	
4.86	-.5339E+02	
4.72	-.4604E+02	
4.58	-.3896E+02	Consistent with
4.44	-.3215E+02	axial velocity
4.31	-.2561E+02	solution
4.17	-.1930E+02	
4.03	-.1307E+02	
3.89	-.6724E+01	
3.75	.2730E-04	
3.61	.7377E+01	
3.47	.1566E+02	
3.33	.2505E+02	
3.19	.3575E+02	
3.06	.4796E+02	
2.92	.6188E+02	
2.78	.7776E+02	
2.64	.9593E+02	
2.50	.1168E+03	

Figure 3-5. Velocity gradient.

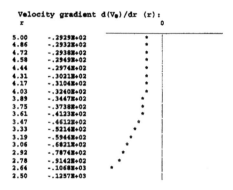

Velocity gradient d(V$_\theta$)/dr (r):

r		0
5.00	-.2929E+02	
4.86	-.2932E+02	
4.72	-.2938E+02	
4.58	-.2949E+02	
4.44	-.2974E+02	
4.31	-.3021E+02	
4.17	-.3104E+02	
4.03	-.3240E+02	
3.89	-.3447E+02	
3.75	-.3738E+02	
3.61	-.4123E+02	
3.47	-.4612E+02	
3.33	-.5214E+02	
3.19	-.5944E+02	
3.06	-.6821E+02	
2.92	-.7874E+02	
2.78	-.9142E+02	
2.64	-.1068E+03	
2.50	-.1257E+03	

Figure 3-6. Velocity gradient.

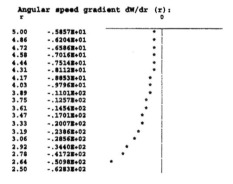

Angular speed gradient dW/dr (r):

r		0
5.00	-.5857E+01	
4.86	-.6204E+01	
4.72	-.6586E+01	
4.58	-.7016E+01	
4.44	-.7514E+01	
4.31	-.8112E+01	
4.17	-.8853E+01	
4.03	-.9796E+01	
3.89	-.1101E+02	
3.75	-.1257E+02	
3.61	-.1456E+02	
3.47	-.1701E+02	
3.33	-.2007E+02	
3.19	-.2386E+02	
3.06	-.2856E+02	
2.92	-.3440E+02	
2.78	-.4172E+02	
2.64	-.5098E+02	
2.50	-.6283E+02	

Figure 3-7. Angular speed gradient.

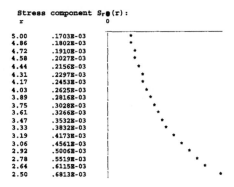

Figure 3-8. Viscous stress.

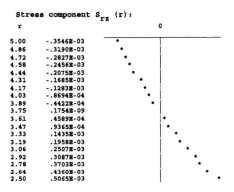

Figure 3-9. Viscous stress.

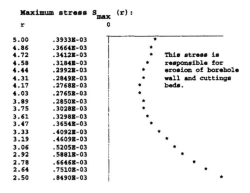

Figure 3-10. Maximum viscous stress.

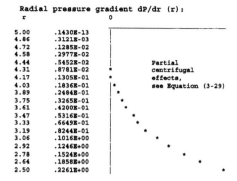

Figure 3-11. Radial pressure gradient.

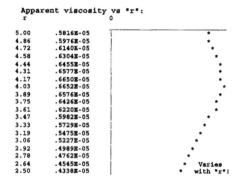

Figure 3-12. Apparent viscosity.

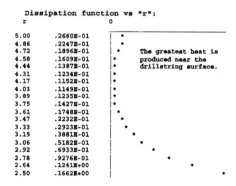

Figure 3-13. Dissipation function.

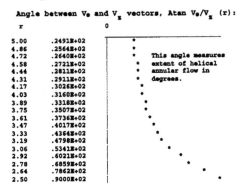

Figure 3-14. Velocity angle.

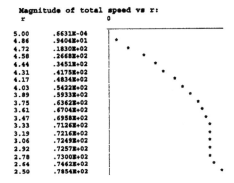

Figure 3-15. Total speed.

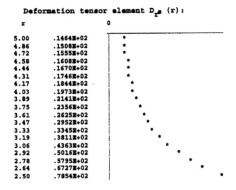

Figure 3-16. Deformation tensor element.

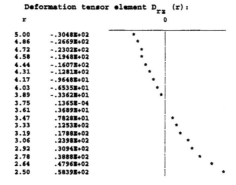

Figure 3-17. Deformation tensor element.

Finally, the computer algorithm calculates radially averaged quantities using the definition

$$F_{avg} = \int_{R_P}^{R_B} F(r)\, dr / (R_B - R_P) \qquad (3\text{-}83)$$

and a second-order accurate integration scheme. Note that this is not a volume weighted average. When properties vary rapidly over r, the linear average (or *any average*) may not be meaningful as a correlation or analysis parameter. Table 3-4 displays computed average results.

Table 3-4
Averaged Values of Annular Quantities

```
Average V_z (in/sec)  =  .3512E+02
            (ft/min)  =  .1756E+03

Average Vθ (in/sec)   =  .3737E+02
Average W (rad/sec)   =  .1160E+02
Average total speed (in/sec) =  .5379E+02

Average angle between V_z and Vθ (deg)  =  .4189E+02
Average d(V_z)/dr (1/sec)  =  .0000E+00

Average d(Vθ)/dr (1/sec)  = -.5028E+02
Average dW/dr (1/(sec X in))  = -.1906E+02
Average dP/dr (psi/in)  =  .5718E-01

Average S_rθ (psi)  =  .3410E-03
Average S_rz (psi)  =  .2432E-04
Average S_max (psi)  =  .4146E-03

Average dissipation function (lbf/(sec sq in))  =  .3753E-01
Average apparent viscosity (lbf sec/sq in)  =  .5876E-05

Average D_rθ (1/sec)  =  .3094E+02
Average D_rz (1/sec)  =  .4445E+01
```

Example 3. More of East Greenbriar

We repeated the calculations for "East Greenbriar No. 2" with all parameters unchanged except for the fluid exponent, which we increased to a near-Newtonian level of 0.9 (again, 1.0 is the Newtonian value). In the first run, we considered a static, nonrotating drillstring with a "rpm" of 0.001, and obtained a volume flow rate of 196.2 gal/min. This is quite different from our earlier 373.6 gal/min, which assumed a fluid exponent of n = 0.724. That is, a 24% increase in the fluid exponent n resulted in a 47% decrease in flow rate; these numbers show how sensitive results are to changes in n. The axial speeds, apparent viscosities, and averaged parameter values obtained are given in Figures 3-18, 3-19, and in Table 3-5. Note how the apparent viscosity is almost constant everywhere with respect to radial position; the well-known localized "pinch" is found near the center of the annulus, where the axial velocity gradient vanishes.

```
Axial speed Vz(r):
  r                      0

 5.00      .8649E-05     |
 4.86      .2948E+01     |    *
 4.72      .6058E+01     |        *
 4.58      .9010E+01     |           *
 4.44      .1171E+02     |             *
 4.31      .1410E+02     |               *
 4.17      .1613E+02     |                 *
 4.03      .1776E+02     |                  *
 3.89      .1897E+02     |                   *
 3.75      .1972E+02     |                    *
 3.61      .1998E+02     |                    *
 3.47      .1971E+02     |                    *
 3.33      .1889E+02     |                   *
 3.19      .1747E+02     |                  *
 3.06      .1541E+02     |                 *
 2.92      .1268E+02     |              *
 2.78      .9239E+01     |           *
 2.64      .5029E+01     |       *
 2.50      .0000E+00     |
```

Figure 3-18. Axial speed.

```
Apparent viscosity vs "r":
  r                       0

 5.00      .1341E-04      |     *
 4.86      .1357E-04      |     *
 4.72      .1375E-04      |     *
 4.58      .1397E-04      |     *
 4.44      .1424E-04      |     *
 4.31      .1457E-04      |     *
 4.17      .1502E-04      |      *
 4.03      .1568E-04      |      *
 3.89      .1690E-04      |        *
 3.75      .4468E-04      |                     *
 3.61      .1683E-04      |        *
 3.47      .1555E-04      |       *
 3.33      .1483E-04      |      *
 3.19      .1433E-04      |     *
 3.06      .1394E-04      |     *
 2.92      .1362E-04      |     *
 2.78      .1335E-04      |     *
 2.64      .1311E-04      |     *
 2.50      .1289E-04      |     *
```

Figure 3-19. Apparent viscosity.

For our second run, we retain the foregoing parameters with the exception of drillstring rpm, which we increase significantly for test purposes from 0.001 to 300 (the fluid exponent is still 0.9). The volume flow rate computed was 232.9 gpm, which is higher than the 196.2 gpm obtained above by a significant 18.7%. Thus, even for "almost Newtonian" power law fluids, the effect of rotation allows a higher flow rate for the same pressure drop. Thus, to produce the lower flow rate, a pump having less pressure output "than normal" would suffice. Computed results are shown in Figures 3-20 and 3-21 and Table 3-6.

Table 3-5
Averaged Values of Annular Quantities

```
Average V_z (in/sec) =  .1305E+02
           (ft/min) =  .6523E+02

Average Vθ (in/sec) =  .3535E-03
Average W (rad/sec) =  .1074E-03
Average total speed (in/sec) =  .1305E+02
Average angle between V_z and V_Õ (deg) =  .2441E+01
Average d(V_z)/dr (1/sec) =  .0000E+00

Average d(Vθ)/dr (1/sec) = -.4288E-03
Average dW/dr (1/(sec X in)) = -.1630E-03
Average dP/dr (psi/in) =  .4719E-11

Average S_rθ (psi) =  .7903E-08
Average S_rz (psi) =  .2432E-04

Average S_max (psi) =  .2088E-03

Average dissipation function (lbf/(sec sq in)) =  .4442E-02
Average apparent viscosity (lbf sec/sq in) =  .1617E-04

Average D_rθ (1/sec) =  .2681E-03
Average D_rz (1/sec) =  .9912E+00
```

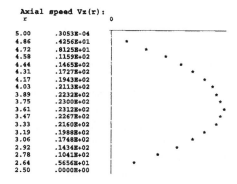

```
Axial speed Vz(r):
  r                    0
5.00      .3053E-04    ┌
4.86      .4256E+01    |      *
4.72      .8125E+01    |        *
4.58      .1159E+02    |          *
4.44      .1465E+02    |            *
4.31      .1727E+02    |             *
4.17      .1943E+02    |              *
4.03      .2113E+02    |               *
3.89      .2232E+02    |                *
3.75      .2300E+02    |                 *
3.61      .2312E+02    |                 *
3.47      .2267E+02    |                 *
3.33      .2160E+02    |                *
3.19      .1988E+02    |               *
3.06      .1748E+02    |              *
2.92      .1434E+02    |            *
2.78      .1041E+02    |          *
2.64      .5656E+01    |       *
2.50      .0000E+00    |
```

Figure 3-20. Axial speed.

```
Apparent viscosity vs "r":
r                         0

5.00      .1294E-04     |                              *
4.86      .1298E-04     |                              *
4.72      .1300E-04     |                              *
4.58      .1300E-04     |                               *
4.44      .1299E-04     |                              *
4.31      .1296E-04     |                              *
4.17      .1291E-04     |                             *
4.03      .1284E-04     |                             *
3.89      .1276E-04     |                            *
3.75      .1266E-04     |                           *
3.61      .1255E-04     |                          *
3.47      .1243E-04     |                         *
3.33      .1231E-04     |                        *
3.19      .1218E-04     |                       *
3.06      .1205E-04     |                      *
2.92      .1191E-04     |                     *
2.78      .1177E-04     |                    *
2.64      .1163E-04     |                   *
2.50      .1148E-04     |                  *
```

Figure 3-21. Apparent viscosity.

Table 3-6
Averaged Values of Annular Quantities

```
Average V_z (in/sec)  =  .1539E+02
            (ft/min)  =  .7693E+02

Average Vθ (in/sec)  =  .3730E+02
Average W (rad/sec)  =  .1162E+02
Average total speed (in/sec)  =  .4150E+02
Average angle between V_z and V_ō (deg)  =  .5993E+02
Average d(V_z)/dr (1/sec)  =  .0000E+00

Average d(Vθ)/dr (1/sec)  =  -.4303E+02
Average dW/dr (1/(sec X in))  =  -.1654E+02
Average dP/dr (psi/in)  =  .5811E-01

Average S_rθ (psi)  =  .6717E-03
Average S_rz (psi)  =  .2432E-04
Average S_max (psi)  =  .7149E-03
Average dissipation function (lbf/(sec sq in))  =  .4851E-01
Average apparent viscosity (lbf sec/sq in)  =  .1251E-04

Average D_rθ (1/sec)  =  .2733E+02
Average D_rz (1/sec)  =  .1350E+01
```

The effect of increasing drillstring rpm has increased the average borehole maximum stress by 3.42 times; this may be of interest to wellbore stability. The apparent viscosity in this example, unlike the previous, is nearly constant everywhere and does not "pinch out." The analytical solutions derived in this chapter are of fundamental rheological interest. But they are particularly useful in drilling and production applications, insofar as the effect of rotation on "volume flow rate versus pressure drop" is concerned, as we will see later in Chapter 5. They allow us to study various operational "what if" questions quickly and efficiently. These solutions also provide a means to correlate experimental data nondimensionally.

REFERENCES

Fredrickson, A.G., and Bird, R.B., "Non-Newtonian Flow in Annuli," *Ind. Eng. Chem.*, Vol. 50, 1958, p. 347.

Savins, J.G., and Wallick, G.C., "Viscosity Profiles, Discharge Rates, Pressures, and Torques for a Rheologically Complex Fluid in a Helical Flow," *A.I.Ch.E. Journal*, Vol. 12, No. 2, March 1966, pp. 357-363.

Schlichting, H., *Boundary Layer Theory*, McGraw-Hill, New York, 1968.

Slattery, J.C., *Momentum, Energy, and Mass Transfer in Continua*, Robert E. Krieger Publishing Company, New York, 1981.

4

Recirculating Annular Vortex Flows

Problems with cuttings accumulation, flow blockage, and resultant stuck pipe in deviated wells are becoming increasingly important operational issues as interest in horizontal drilling continues. For small angles (ß) from the vertical, annular flows and hole cleaning are well understood; for example, cleaning efficiency is always improved by increasing velocity, viscosity, or both. But beyond 30 degrees, these issues are rife with challenging questions. Many unexplained, confusing, and conflicting observations are reported by different investigators; however, it turns out that bottom viscous stress (which tends to erode cuttings beds having well-defined yield stresses) is the correlation parameter that explains many of these discrepancies.

The next chapter addresses problems related to cuttings transport using the eccentric and rotating flow models developed earlier. Cuttings accumulation, of course, is dangerous because the resulting blockage of the annular space increases the possibility of stuck pipe. Here, we will learn that *flows can be blocked even when there is no externally introduced debris* in the system. In other words, *dangerous flow blockage can arise from fluid-dynamical effects alone.* This possibility is very real whenever there exist density gradients in a direction perpendicular to the flow, e.g., "barite sag" in the context of drilling. This blockage is also possible in pipe flows of slurries, for instance, slurries carrying ground wax and hydrate particles or other debris.

In Chapters 2 and 3, the pressure field was assumed to be uniform across the annulus; the velocity field is therefore unidirectional, with the fluid flowing axially from high pressure regions to low. These assumptions are reasonable since numerous flows do behave in this manner. In this chapter, we will relax these assumptions, but turn to more general flows with density stratification. A special class of annular and pipe flow lends itself to strange occurrences we call "recirculating vortex flows," to which we now turn our attention.

WHAT ARE RECIRCULATING VORTEX FLOWS?

In deviated holes where circulation has been temporarily interrupted, weighting material such as barite, drilled cuttings or cement additives, may fall out of suspension. Similarly, pipelines containing slurries with wax or hydrate particles can develop vertical density gradients when flow is temporarily slowed or halted. This gravity stratification has mass density increasing downwards. And this stable stratification, which we collectively refer to as "barite sag," is thought to be responsible for the trapped, self-contained "recirculation zones" or "bubbles" observed by many experimenters.

They contain rotating, swirling, "ferris-wheel-like" motions within their interiors; the external fluid that flows around them "sees" these zones as stationary obstacles that impede their axial movement up the annulus or pipe. Excellent color video tapes showing these vortex-like motions in detail have been produced by M-I Drilling Fluids, which were viewed by the author in its Houston facilities prior to the initial printing of *Borehole Flow Modeling*.

These strange occurrences are just that; their appearances seem to be sporadic and unpredictable, as much myth as reality. However, once they are formed, they remain as stable fluid-dynamical structures that are extremely difficult to remove. They are dangerous and undesirable because of their tendency to entrain cuttings, block axial flow, and increase the possibility of stuck pipe. One might ask, "Why do these bubbles form? What are the controlling parameters? How can their occurrences be prevented?"

Detailed study of M-I's tapes suggests that the recirculating flows form independently of viscosity and rheology to leading order, that is, they do not depend primarily on "n" and "k." They appear to be inertia-dominated, depending on density effects themselves, while nonconservative viscous terms play only a minor role in sustaining or damping the motion. This leaves the component of density stratification normal to the hole axis as the primary culprit; it alone is responsible for the highly three-dimensional pressure field that drives local pockets of secondary flow. It is possible, of course, to have multiple bubbles coexisting along a long deviated hole.

Again these recirculating bubbles, observed near pipe bends, stabilizers, and possibly marine risers and other obstructions, are important for various practical reasons. First, they block the streamwise axial flow, resulting in the need for increased pressure to pass a given volume flow rate. Second, because they entrain the mud and further trap drilled cuttings, they are a likely cause of stuck pipe. Third, the external flow modified by these bubbles can also affect the very process of cuttings bed formation and removal itself.

Fortunately, these bubbles can be studied, modeled, and characterized in a rather simple manner; very instructive "snapshots" of streamline patterns covering a range of vortex effects are given later. This chapter identifies the *nondimensional* channel parameter **Ch** responsible for vortex bubble formation and describes the physics of these recirculating flows. The equations of motion are given and solved using finite difference methods for several practical flows. The detailed bubble development process is described and illustrated in a sequence of computer-generated pictures.

MOTIVATING IDEAS AND CONTROLLING VARIABLES

The governing momentum equations are Euler's equations, which describe large-amplitude, inviscid shear flow in both stratified and unstratified media (Schlichting, 1968; Turner, 1973). For the problem at hand, they simplify to

$$\rho \{uu_x + vu_y\} = -p_x \qquad (4\text{-}1)$$

$$\rho \{uv_x + vv_y\} = -p_y - g\rho \cos \alpha \qquad (4\text{-}2)$$

$$\{\rho u\}_x + \{\rho v\}_y = 0 \qquad (4\text{-}3)$$

We have assumed that the steady vortex flow is contained in a two-dimensional rectangular box in the plane of the hole axis (x) and the direction of density stratification (y). This is based on experimental observation: the vortical flows do not wrap around the drillpipe. In the above equations, u and v are velocities in the x and y directions, respectively. Subscripts indicate partial derivatives. Also, ρ is a fluid density that varies linearly with y far upstream, $p(x,y)$ is the unknown pressure field, g is the acceleration due to gravity, and α is the angle the borehole axis makes with the horizontal ($\alpha + \beta = 90^o$; also see Chapters 2 and 3). Equations 4-1 and 4-2 are momentum equations in the x and y directions, while Equation 4-3 describes mass conservation.

Nondimensional parameters are important to *understanding* physical events. The well-known Reynolds number, which measures the relative effects of inertia to viscous forces, is one example of a nondimensional parameter. It alone, for example, dictates the onset of turbulence; also, like Reynolds numbers imply dynamically similar flow patterns. Analogous nondimensional variables are used in different areas of physics; for instance, the mobility ratio in reservoir engineering or the Mach number in high-speed aerodynamics.

Close examination of Equations 4-1 to 4-3 using affine transformations shows that the physics of bubble formation depends on a *single* nondimensional variable **Ch** characterizing the channel flow. It is constructed from the combination of two simpler ones. The first is a Froude number $U^2/gL \cos \alpha$, where U is the average oncoming speed and L is the channel height between the pipe and borehole walls. The second is a relative measure of stratification, say $d\rho/\rho_{ref}$ ($d\rho$ might represent the density difference between the bottom and top of the annulus or pipe, and ρ_{ref} may be taken as their arithmetic average). The combined parameter **Ch** of practical significance is

$$Ch = U^2 \rho_{ref}/gL \, d\rho \cos \alpha \qquad (4\text{-}4)$$

We now summarize our findings. For large values of **Ch**, recirculation bubbles will *not* form; the streamlines of motion are essentially straight and the rheology-dominated models developed in Chapters 2 and 3 apply. For small **Ch**'s of order unity, small recirculation zones *do* form, and elongate in the streamwise dimension as **Ch** decreases. For still smaller values, that is, values below a critical value of 0.3183, solutions with wavy upstream flows are found, which may or may not be physically realistic.

Equations 4-1 to 4-3 can be solved using "brute force" computational methods, but they are more cleverly treated by introducing the streamfunction used by aerodynamicists and reservoir engineers. When the problem is reformulated in this manner, the result is a nonlinear Poisson equation that can be easily integrated using fast iterative solvers. Streamlines are obtained by connecting computed streamfunction elevations having like values. The arithmetic difference in streamfunction between any two points is directly proportional to the volume flow rate passing through the two points. Velocity and pressure fields can be obtained by post-processing the computed streamfunction solutions computed.

DETAILED CALCULATED RESULTS

Typically, the solution obtained for a 20×40 mesh will require less than one second on Pentium machines. In Figures 4-1 to 4-6, we have allowed the flow to "disappear" into a "mathematical sink" (in practice, the distance to this obstacle over the height L will appear as a second ratio). This sink simulates the presence of obstacles or pipe elbows located further upstream. With decreasing values of **Ch**, the appearance of an elongating recirculation bubble is seen. The computed streamline patterns, again constructed by drawing level contours, depict an ever-worsening flowfield. Each figure displays "raw" streamfunction data as well as processed contour plots.

The stand-alone vortexes so obtained are inherently stable, since they represent patches of angular momentum that physical laws insist must be conserved. In this sense, they are not unlike isolated trailing aircraft tip vortices, that persist indefinitely in the air until dissipation renders them harmless. However, annular bubbles are worse: the channel flow itself is what drives them, perpetuates them, and increases their ability to do harm by further entraining solid debris in the annulus. In the pipeline context, slurry density gradients likewise promote flow blockage and vortex recirculation; the resulting "sandpapering," allowing continuous rubbing against pipe walls, can lead to metal erosion, decreased strength, and unexpected rupture.

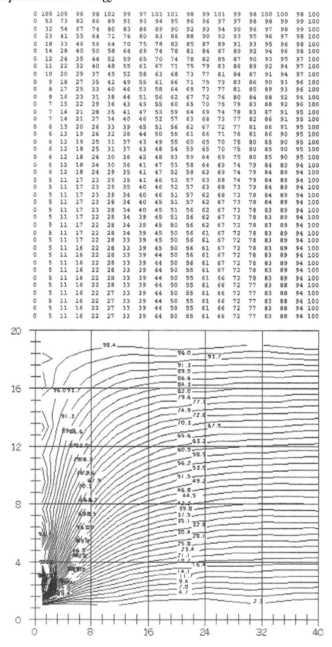

Figure 4-1. Ch = 1.0, straight streamlines without recirculation.

```
0 105 105  96  98 102  99  97 101 101  98  99 101  99  98 100 100  98 100
0  54  74  83  88  91  93  94  96  96  97  98  98  98  99  99  99  99 100
0  33  55  68  77  82  86  89  91  93  94  95  96  97  98  98  99  99 100
0  24  43  57  67  74  80  84  87  89  91  93  94  96  97  97  98  99 100
0  19  35  49  59  67  74  79  83  86  88  91  92  94  95  97  98  99 100
0  15  30  43  53  61  68  74  79  82  86  88  90  92  94  96  97  98 100
0  13  26  38  48  56  64  70  75  79  83  86  88  91  93  95  96  98 100
0  12  23  34  44  52  60  66  71  76  80  83  86  89  92  94  96  98 100
0  11  21  31  41  49  56  63  68  73  77  81  85  88  90  93  95  97 100
0  10  20  29  38  46  53  60  65  71  75  79  83  86  89  92  95  97 100
0   9  18  27  36  43  51  57  63  68  73  77  81  85  88  91  94  97 100
0   8  17  26  34  41  48  55  61  66  71  76  80  83  87  90  93  97 100
0   8  16  24  32  40  46  53  59  64  69  74  78  82  86  90  93  96 100
0   8  16  23  31  38  45  51  57  63  68  73  77  81  85  89  93  96 100
0   7  15  22  30  37  43  50  56  61  66  71  76  80  84  88  92  96 100
0   7  14  22  29  36  42  48  54  60  65  70  75  79  84  88  92  96 100
0   7  14  21  28  35  41  47  53  59  64  69  74  78  83  87  91  95 100
0   7  14  20  27  34  40  46  52  58  63  68  73  78  82  87  91  95 100
0   6  13  20  26  33  39  45  51  57  62  67  72  77  82  86  91  95 100
0   6  13  19  26  32  38  44  50  56  61  67  72  76  81  86  90  95 100
0   6  13  19  25  32  38  44  50  55  61  66  71  76  81  86  90  95 100
0   6  12  19  25  31  37  43  49  55  60  65  70  75  80  85  90  95 100
0   6  12  18  25  31  37  43  48  54  59  65  70  75  80  85  90  95 100
0   6  12  18  24  30  36  42  48  53  59  64  70  75  80  85  90  95 100
0   6  12  18  24  30  36  42  47  53  58  64  69  74  79  85  90  95 100
0   6  12  18  24  30  36  41  47  53  58  63  69  74  79  84  89  94 100
0   6  12  18  23  29  35  41  47  52  58  63  69  74  79  84  89  94 100
0   5  11  17  23  29  35  40  46  52  57  63  68  73  79  84  89  94 100
0   5  11  17  23  29  35  40  46  52  57  63  68  73  79  84  89  94 100
0   5  11  17  23  29  34  40  46  51  57  62  68  73  78  84  89  94 100
0   5  11  17  23  29  34  40  46  51  57  62  68  73  78  84  89  94 100
0   5  11  17  23  28  34  40  45  51  57  62  67  73  78  84  89  94 100
0   5  11  17  23  28  34  40  45  51  56  62  67  73  78  83  89  94 100
0   5  11  17  22  28  34  39  45  51  56  62  67  73  78  83  89  94 100
0   5  11  17  22  28  34  39  45  51  56  62  67  73  78  83  89  94 100
0   5  11  17  22  28  34  39  45  50  56  61  67  72  78  83  89  94 100
0   5  11  17  22  28  34  39  45  50  56  61  67  72  78  83  89  94 100
0   5  11  17  22  28  33  39  45  50  56  61  67  72  78  83  89  94 100
0   5  11  16  22  28  33  39  45  50  56  61  67  72  78  83  89  94 100
0   5  11  16  22  28  33  39  45  50  56  61  67  72  78  83  89  94 100
0   5  11  16  22  28  33  39  44  50  56  61  67  72  78  83  89  94 100
```

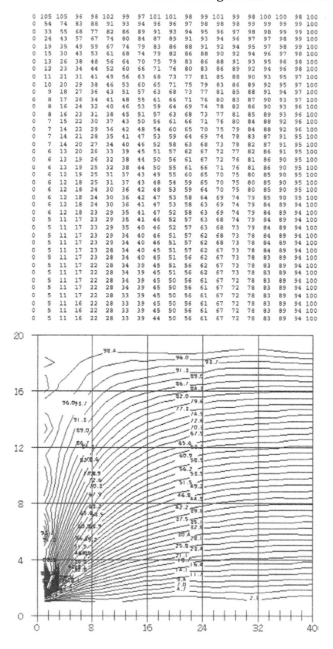

Figure 4-2. Ch = 0.5, straight streamlines without recirculation.

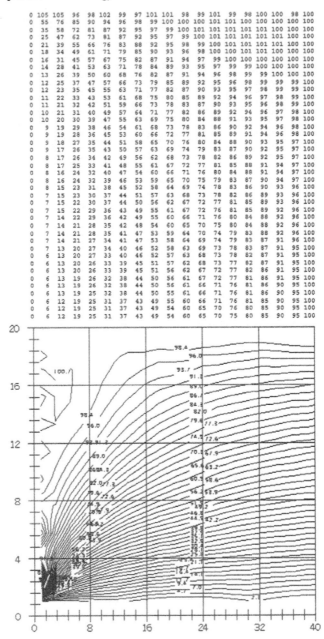

```
0 105 105  96  98 102  99  97 101 101  98  99 101  99  98 100 100  98 100
0  55  76  85  90  94  96  98  99 100 100 100 101 101 100 100 100 100 100
0  35  58  72  81  87  92  95  97  99 100 101 101 101 101 101 100 100 100
0  25  47  62  73  81  87  92  95  97  99 100 101 101 101 101 101 100 100
0  21  39  55  66  76  83  88  92  95  98  99 100 101 101 101 101 100 100
0  18  34  49  61  71  79  85  90  93  96  98 100 100 101 101 101 100 100
0  16  31  45  57  67  75  82  87  91  94  97  99 100 100 100 100 100 100
0  14  28  41  53  63  71  78  84  89  93  95  97  99  99 100 100 100 100
0  13  26  39  50  60  68  76  82  87  91  94  96  98  99  99 100 100 100
0  12  25  37  47  57  66  73  79  85  89  92  95  96  98  99  99  99 100
0  12  23  35  45  55  63  71  77  82  87  90  93  95  97  98  99  99 100
0  11  22  33  43  53  61  68  75  80  85  89  92  94  96  97  98  99 100
0  11  21  32  42  51  59  66  73  78  83  87  90  93  95  96  98  99 100
0  10  21  31  40  49  57  64  71  77  82  86  89  92  94  96  97  98 100
0  10  20  30  39  47  55  63  69  75  80  84  88  91  93  95  97  98 100
0   9  19  29  38  46  54  61  68  73  78  83  86  90  92  94  96  98 100
0   9  19  28  36  45  53  60  66  72  77  81  85  89  91  94  96  98 100
0   9  18  27  35  44  51  58  65  70  76  80  84  88  90  93  95  97 100
0   9  17  26  35  43  50  57  63  69  74  79  83  87  90  92  95  97 100
0   8  17  26  34  42  49  56  62  68  73  78  82  86  89  92  95  97 100
0   8  17  25  33  41  48  55  61  67  72  77  81  85  88  91  94  97 100
0   8  16  24  32  40  47  54  60  66  71  76  80  84  88  91  94  97 100
0   8  16  24  32  39  46  53  59  65  70  75  79  83  87  90  93  96 100
0   8  15  23  30  37  44  51  57  63  68  73  78  82  86  89  93  96 100
0   7  15  23  30  37  44  50  56  62  67  72  77  81  85  89  93  96 100
0   7  15  22  29  36  43  49  55  61  67  72  76  81  85  89  92  96 100
0   7  14  22  29  36  42  49  55  60  66  71  76  80  84  88  92  96 100
0   7  14  21  28  35  42  48  54  60  65  70  75  80  84  88  92  96 100
0   7  14  21  28  35  41  47  53  59  64  70  74  79  83  88  92  96 100
0   7  14  21  27  34  41  47  53  58  64  69  74  79  83  87  91  96 100
0   7  15  22  30  37  44  50  56  62  67  72  77  81  85  89  93  96 100
0   7  13  20  27  34  40  46  52  58  63  69  73  78  83  87  91  95 100
0   6  13  20  27  33  40  46  52  57  63  68  73  78  82  87  91  95 100
0   6  13  20  26  33  39  45  51  57  62  68  73  77  82  87  91  95 100
0   6  13  19  26  32  38  44  50  56  61  67  72  77  81  86  91  95 100
0   6  13  19  26  32  38  44  50  56  61  66  71  76  81  86  90  95 100
0   6  13  19  25  32  38  44  50  55  61  66  71  76  81  86  90  95 100
0   6  12  19  25  31  37  43  49  55  60  66  71  76  81  85  90  95 100
0   6  12  19  25  31  37  43  49  54  60  65  70  76  80  85  90  95 100
0   6  12  19  25  31  37  43  49  54  60  65  70  75  80  85  90  95 100
```

Figure 4-3. Ch = 0.35, minor recirculating vortex.

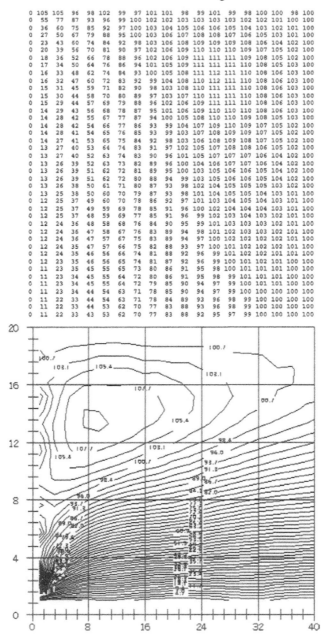

Figure 4-4. Ch = 0.320, large scale recirculation.

```
0 105 105  96  98 102  99  97 101 101  98  99 101  99  98 100 100  98 100
0  55  77  87  93  97  99 101 102 103 103 103 103 103 102 102 101 100 100
0  36  61  76  86  93  98 101 104 105 106 106 106 106 105 104 102 101 100
0  27  50  67  80  89  96 101 104 107 108 109 109 108 107 105 104 102 100
0  23  44  61  75  85  93 100 104 107 109 110 111 110 109 107 105 102 100
0  20  40  57  71  82  91  99 104 108 110 112 112 111 110 108 105 102 100
0  19  37  53  68  80  90  97 104 108 111 112 113 112 111 109 106 103 100
0  18  35  51  65  78  88  96 103 108 111 113 114 113 111 109 106 103 100
0  17  34  49  63  76  87  95 102 108 111 113 114 113 112 110 107 103 100
0  16  33  48  62  75  85  94 102 107 111 113 114 114 112 110 107 103 100
0  16  32  47  61  73  84  94 101 107 111 113 114 114 112 110 107 103 100
0  16  31  46  60  72  83  93 100 106 110 113 114 114 112 110 107 103 100
0  15  31  45  59  72  83  92 100 106 110 113 114 114 112 110 107 103 100
0  15  30  45  58  71  82  91  99 105 110 112 114 113 112 110 107 103 100
0  15  30  44  58  70  81  91  98 105 109 112 113 113 112 110 107 103 100
0  15  30  44  57  70  81  90  98 104 109 111 113 113 112 110 107 103 100
0  15  29  44  57  69  80  89  97 103 108 111 113 113 112 109 107 103 100
0  14  29  43  56  69  79  89  97 103 108 111 112 112 111 109 106 103 100
0  14  29  43  56  68  79  88  96 102 107 110 112 112 111 109 106 103 100
0  14  29  42  56  68  78  88  96 102 107 110 111 112 111 109 106 103 100
0  14  28  42  55  67  78  87  95 101 106 109 111 111 110 109 106 103 100
0  14  28  42  55  67  77  87  94 101 105 109 110 111 110 108 106 103 100
0  14  28  42  54  66  77  86  94 100 105 108 110 111 110 108 106 103 100
0  14  28  41  54  66  76  86  93 100 104 108 110 110 109 108 106 103 100
0  14  28  41  54  65  76  85  93  99 104 107 109 110 109 108 105 103 100
0  14  27  41  53  65  75  85  92  99 103 107 109 109 109 107 105 102 100
0  14  27  41  53  65  75  84  92  98 103 106 108 109 109 107 105 102 100
0  13  27  40  53  64  75  84  91  98 102 106 108 109 108 107 105 102 100
0  13  27  40  52  64  74  83  91  97 102 105 107 108 108 107 105 102 100
0  13  27  40  52  63  74  83  90  97 102 104 107 108 108 106 105 102 100
0  13  27  40  52  63  73  82  90  96 101 104 107 107 107 106 104 102 100
0  13  26  39  51  63  73  82  89  96 101 104 106 107 107 106 104 102 100
0  13  26  39  51  62  72  81  89  95 100 104 106 107 107 106 104 102 100
0  13  26  39  51  62  72  81  89  95 100 103 105 106 106 105 104 102 100
0  13  26  39  51  62  72  81  88  94  99 103 105 106 106 105 104 102 100
0  13  26  38  50  61  71  80  88  94  99 102 105 106 106 105 104 102 100
0  13  26  38  50  61  71  80  87  93  98 102 104 105 105 105 103 102 100
0  13  25  38  50  61  70  79  87  93  98 101 104 105 105 105 103 102 100
0  13  25  38  49  60  70  79  86  93  97 101 103 105 105 104 103 101 100
0  12  25  37  49  60  70  78  86  92  97 101 103 104 105 104 103 101 100
0  12  25  37  49  60  69  78  85  92  97 100 103 104 104 104 103 101 100
```

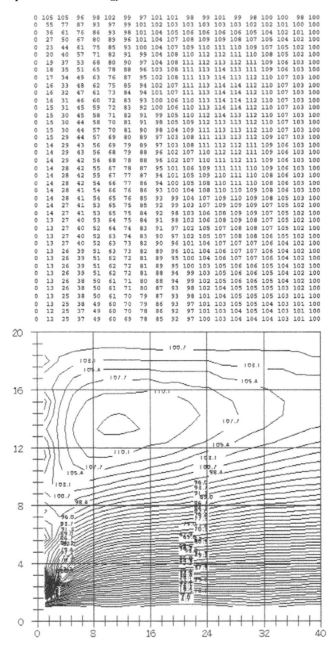

Figure 4-5. Ch = 0.319, major flow blockage.

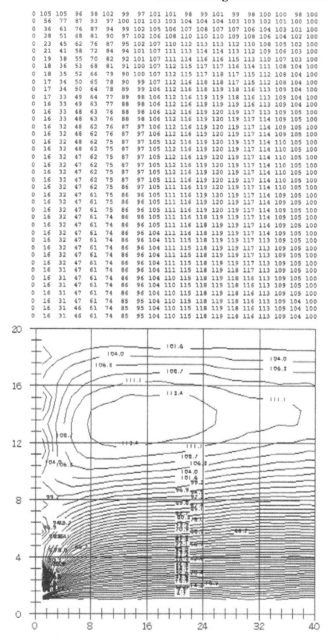

Figure 4-6. Ch = 0.3185, major flow blockage by elongated vortex structure.

HOW TO AVOID STAGNANT BUBBLES

We have shown that recirculating zones can develop from interactions between inertia and gravity forces. These bubbles form when density stratification, hole deviation and pump rate fulfill certain special conditions. These are elegantly captured in a single channel variable, the nondimensional parameter $Ch = U^2 \rho_{ref}/gLd\rho \cos \alpha$. Moreover, the resulting flowfields can be efficiently computed and displayed, thus allowing us to understand better their dynamical consequences.

Suppressing recirculating flows is simply accomplished: *avoid small values of the nondimensional Ch parameter*. Small values, as is evident from Equation 4-4, can result from different isolated effects. For example, it decreases as the hole becomes more horizontal, as density differences become more pronounced, or as pumping rates decrease. But none of these factors alone control the physics; it is the combination *taken together* that controls bubble formation and perhaps the fate of a drilling program.

We have modeled the problem as the single phase flow of a stratified fluid, rather than as the combined motion of dual-phase fluid and solid continuum. This simplifies the mathematical issues without sacrificing the essential physical details. For practical purposes, the parameter Ch can be viewed as a "danger indicator" signaling impending cuttings transport or stuck pipe problems. It is the single most important parameter whenever interrupted circulation or poor suspension properties lead to gravity segregation and settling of weighting materials in drilling mud.

These considerations also apply to cementing, where density segregation due to gravity *and* slow velocities are both likely. When recirculation zones form in either the mud or the cement above or beneath the casing, the displacement effectiveness of the cement is severely impeded. The result is mud left in place, an undesirable one necessitating squeeze jobs. Similar remarks apply to pipeline applications. Recirculation zones are likely to be encountered at low flow rates that promote density stratification, and immediately prior to flow start-up, when slurry particles have been allowed to settle out.

We emphasize that the vortical bubbles considered here are *not* the "Taylor vortices" studied in the classical fluid mechanics of homogeneous flows. Taylor vortices are "doughnuts" that would normally "wrap around," in our case, the drillpipe; to the author's knowledge, these have not been observed in drilling applications. They can be created in the absence of density stratification, that is, they can be found in purely homogeneous fluids. Importantly, Taylor vortices would owe their existence to finite drillstring length effects, and would represent completely different physical mechanisms.

A PRACTICAL EXAMPLE

We have discussed the dynamical significance of the nondimensional parameter **Ch** that appears in the normalized equations of motion. For use in practical estimates, the channel variable may be written more clearly as a multiplicative sequence of dimensionless entities,

$$\mathbf{Ch} = U^2 \rho_{ref}/gLd\rho \cos \alpha \qquad (4\text{-}5)$$

$$= (U^2/gL) \times (\rho_{ref}/ d\rho) \times (1/\cos \alpha)$$

Let us consider an annular flow studied in the cuttings transport examples of Chapter 5. For the 2-inch and 5-inch pipe and borehole radii, the cross-sectional area is $\pi (5^2 - 2^2)$ or 66 in^2.

The experimental data used in Discussions 1 and 2 of Chapter 5 assume oncoming linear velocities of 1.91, 2.86 and 3.82 ft/sec. Since 1 ft/sec corresponds to a volume flow rate of 1 ft/sec $\times$ 66 in^2 or 205.7 gpm, the flow rates are 393, 588 and 786 gpm. So, at the lowest flow rate of 393 gpm (a reasonable field number), the average linear speed over the entire annulus is approximately 2 ft/sec. But the low-side average will be much smaller, say 0.5 ft/sec. And if the pipe is displaced halfway down, the length scale L will be roughly (5-2) /2 inches or 0.13 ft.

Thus, the first factor in Equation 4-5 takes the value $U^2/gL = (0.5)^2/(32.2 \times 0.13) = 0.06$. If we assume a 20% density stratification, then $\rho_{ref}/d\rho = 5.0$; the product of the two factors is 0.30. For a highly deviated well inclined 70° from the vertical axis, $\alpha = 90° - 70° = 20°$ and $\cos 20° = 0.94$. Thus, we obtain **Ch** = 0.30/0.94 = 0.32. This value, as Figures 4-1 to 4-6 show, lies just at the threshold of danger. Velocities lower than the assumed value are even more likely to sustain recirculatory flows; higher ones, in contrast, are safer.

Of course, the numbers used above are only estimates; a three-dimensional, viscous solution is required to establish true length and velocity scales. But these approximate results show that bottomhole conditions typical of those used in drilling and cementing *are* associated with low values of **Ch** near unity.

We emphasize that **Ch** is the only nondimensional parameter appearing in Equations 4-1 to 4-3. Another one describing the geometry of the annular domain would normally appear through boundary conditions. For convenience though (and for the sake of argument only), we have replaced this requirement with an idealized "sink." In any real calculation, exact geometrical effects must be included to complete the formulation. Also note that our recirculating flows get worse as the borehole becomes more horizontal; that is, **Ch** decreases as α becomes smaller. This is in stark contrast to the unidirectional, homogeneous flows of Chapter 5, which, as we will prove, perform worst near 45°, at least with respect to cuttings transport efficiency. The structure of Equation 4-4 correctly shows that in near-vertical wells with α approaching 90°, **Ch** tends to infinity; thus, the effects of flow blockage due to the vortical bubbles considered here are relegated to highly deviated wells.

Again, flow properties such as local velocity, shear rate, and pressure can be obtained from the computed streamfunction straightforwardly. They may be useful correlation parameters for cuttings transport efficiency and local bed buildup. Continuing research is underway, exploring similarities between this problem and the density-dependent flows studied in dynamic meteorology and oceanography. Obvious extensions of our observations for annular flow apply to the pipeline transport of wax and hydrate slurries.

REFERENCES

Schlichting, H., *Boundary Layer Theory*, McGraw-Hill, New York, 1968.

Turner, J.S., *Buoyancy Effects in Fluids*, Cambridge University Press, London, 1973.

5

Applications to Drilling and Production

In Chapters 2, 3, and 4, we formulated and solved three distinct annular flow models and gave numerous calculated results. These models remove many of the restrictions usually made regarding eccentricity, rotation, and flow homogeneity. The methods also produce fast and stable solutions, with highly detailed output processing. Unfortunately, it is not yet possible to build a single model for rotating, inhomogeneous flow in eccentric domains. This chapter deals with practical applications of these models in drilling and production. Topics discussed include cuttings transport in deviated holes, stuck pipe evaluation, cementing, and coiled tubing return flow analysis. For the casual reader, brief summaries of the three models are offered first.

Recapitulation. Let us briefly summarize the modeling capabilities developed. The first annular flow model applies to homogeneous Newtonian, power law, Bingham plastic, and Herschel-Bulkley fluids. It assumes a nonrotating drillpipe (or casing) in an inclined hole, but no restrictions are placed on the annular geometry itself. The unidirectional analysis handles eccentric circular drillpipes and boreholes; but it also models, with no extra difficulty or computational expense, deformed borehole walls due to erosion, swelling or cuttings bed buildup, and, for example, square drill collars with stabilizers. A description of this capability was first offered in Chin (1990a).

The exact PDEs are solved on boundary conforming grid systems that yield high resolution in tight spaces. They handle geometry exactly, so that slot flow and bipolar coordinate approximations are unnecessary. The unconditionally stable finite difference program, requiring five seconds per simulation on Pentium machines, has been successfully executed thousands of times without diverging since the original publication.

To help understand the lengthy output, which includes annular velocity, apparent viscosity, two components of viscous stress and shear rate, Stokes product and dissipation function, a special character-based graphics program was developed. This utility overlays results on the annular geometry itself, thus facilitating physical interpretation and visual correlation of computed quantities with annular position. The algorithms and graphics software are written in standard Fortran; they are easily ported to most hardware platforms with minimal modification. Calculations and displays for several difficult geometries are given in Chapter 2. With the present revision, more sophisticated color graphics capabilities are available, but the original text diagrams still provide the "quantitative feeling" necessary in understanding engineering problems.

The second flow model provides approximate analytical solutions for power law fluids in concentric annular flows containing rotating drillstrings or casings. Here, a "narrow annulus" assumption is invoked; hole inclination, as in the first model, scales out by using a normalized pressure. Closed form solutions are derived for axial and circumferential velocity, pressure, viscous stress, deformation rate, apparent viscosity and local heat generation. All the properties listed are given as explicit functions of the radial coordinate "r." Again, built-in, portable graphical utilities permit quick, convenient displays of all relevant flow parameters. The corresponding solutions for Newtonian flow are also given in closed analytical form, but no restrictions on annular geometry are required; here, the solutions satisfy the Navier-Stokes equations exactly.

The third and final model considers density heterogeneities due to "barite sag," or, more generally, gravity stratification. This opens up the possibility of reversed flow in the *axial* direction, and annular blockage of an entirely fluid-dynamical nature. This recirculating flow is to be contrasted with better known secondary vortex flows that occur in the *cross-sectional* planes of pipes. Unlike the latter, which are controlled by viscosity, analysis shows that the existence of barite "trapped bubbles" depends on a single nondimensional parameter that is independent of the stress versus strain relation.

The fluid density "ρ" does not explicitly appear in the momentum equations for the first two models, because geometrical simplifications eliminate the convective term associated with "$F = ma$." The influence of "mud weight" appears indirectly, only through its rheological effect on n, k, and yield strength. On the other hand, rheology (to be precise, the exact stress versus strain relation) does not play a fundamental role in sag-induced bubbles; flow reversals due to gravity segregation are primarily inertia dominated. Of course, flows can be found that run counter to our assumptions; however, the author believes that the models chosen for exposition correctly describe physical problems in actual drilling and production.

CUTTINGS TRANSPORT IN DEVIATED WELLS

Recent industry interest in horizontal and highly deviated wells has heightened the importance of annular flow modeling as it relates to hole cleaning. Cuttings transport to the surface is generally impeded by virtue of hole orientation; this is worsened by decreased "low-side" annular velocities due to pipe eccentricity. In addition, the blockage created by bed buildup decreases overall flow rate, further reducing cleaning efficiency. In what could possibly be a self-sustaining, destabilizing process, stuck pipe is a likely end result. This section discloses new cuttings transport correlations and suggests simple predictive measures to avoid bed buildup. Good hole cleaning and bed removal, of course, are important to cementing as well.

Few useful annular flow models are available despite their practical importance. The nonlinear equations governing power law viscous fluids, for example, must be solved with difficult no-slip conditions for highly eccentric geometries. Recent slot flow models offer some improvement over parallel plate approaches. Still, because they unrealistically require slow radial variations in the circumferential direction, large errors are possible. Even when they apply, these models can be cumbersome; they involve "elliptic integrals," which are too awkward for field use. Bipolar coordinate models accurately simulate eccentric flows with circular pipes and boreholes; however, they cannot be extended to real world applications containing washouts and cuttings beds.

In this section, the eccentric flow model is used to interpret field and laboratory results. Because the model actually simulates reality, it has been possible to correlate problems associated with cuttings transport and stuck pipe to unique average mechanical properties of the computed flowfield. These correlations are discussed next.

Discussion 1. Water-Base Muds

Detailed computations using the eccentric model are described, assuming a power law fluid, which correspond to the comprehensive suite of cuttings transport experiments conducted at the University of Tulsa (Becker, Azar and Okrajni, 1989). For a fixed inclination and oncoming flow rate, we importantly demonstrate that "cuttings concentration" correlates linearly with *the mean viscous shear stress averaged over the lower half of the annulus.* Thus, impending cuttings problems can be eased by first determining the existing average stress level; and then, adjusting n, k, and gpm values to increase that stress. Physical arguments supporting our correlations will be given. We emphasize that the present approach is completely predictive and deterministic;

it does *not* require empirical assumptions related to the "equivalent hydraulic radius" or to questionable "pipe to annulus conversion factors." This result was first reported in Chin (1990b), and the "stress correlation" is now used in several service company models for cuttings transport application.

Detailed experimental results for cuttings concentration, a useful indicator of transport efficiency and carrying capacity, were obtained at the University of Tulsa's large scale flow loop. Fifteen bentonite-polymer, water-based muds, for three average flow rates (1.91, 2.86 and 3.82 ft/sec), at three borehole inclinations from vertical (30, 45 and 70 deg), were tested. Table 1 of Becker *et al.* (1989) summarizes all measured mud properties, along with specific power law exponents n and consistency factors k. We emphasize that "water base" does *not* imply Newtonian flow; in fact, the reported values of n differ substantially from unity. The annular geometry consisted of a 2-inch-radius pipe, displaced downward by 1.5 inches in a 5-inch-radius borehole; also, the pipe rotated at 50 rpm.

With flow rate and hole inclination fixed, the authors cross-plot the nondimensional cuttings concentration C versus particular rheological properties for each mud type used. These include apparent viscosity, plastic viscosity (PV), yield point (YP), YP/PV, initial and ten-minute gel strength, "effective viscosity," k, and Fann dial readings at various rpms. Typically, the correlations obtained were poor, with one exception to be discussed. That good correlations were not possible, of course, is not surprising; the "fluid properties" in Becker *et al.* (1989) are rotational viscometer readings describing the test instrument only. That is, they have no real bearing to the actual annular geometry and its downhole flow, which from Chapter 2 we understand can be complicated.

These cross-plots and tables, numbering over 20, were nevertheless studied in detail; using them, the entire laboratory database was reconstructed. The concentric, rotating flow program described in Chapter 3 was run to show that rpm effects were likely to be insignificant. The eccentric annular model of Chapter 2 was then executed for each of the 135 experimental points; detailed results for *calculated* apparent viscosity, shear rate, viscous stress and axial velocity, all of which varied spatially, were tabulated and statistically analyzed along with the experimental data.

Numerous cross-plots were produced, examined, and interpreted. The most meaningful correlation parameter found was the *mean viscous shear stress*, obtained by averaging computed values over the bottom half of the annulus, where cuttings in directional wells are known to form beds. Figures 5-1, 5-2, and 5-3 display cuttings concentration versus our mean shear stress for different average flow speeds and inclination angles ß from the vertical. Each plotted symbol represents a distinct test mud. Calculated correlation coefficients averaged a high 0.91 value. Our correlations apply to laminar flow only; the flattened portions of the curves refer to turbulent conditions. Computationally, the latter are simulated with a subroutine change to include turbulence modeling.

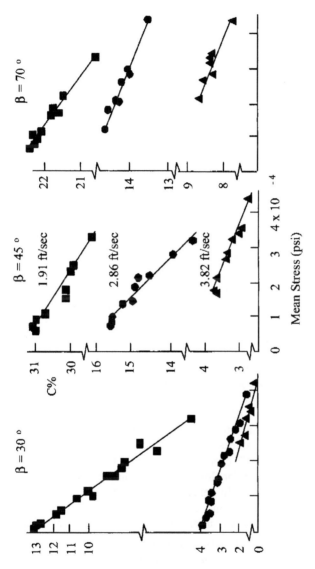

Figures 5-1, 2, 3. Cuttings transport correlation.

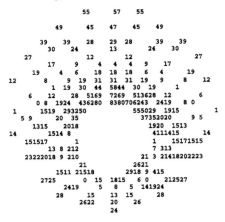

Figure 5-4. Viscous stress.

The program produces easily understood information. Figure 5-4 displays, for example, calculated areal results for viscous shear stress in the visual format described earlier. Tabulated results, in this case for "Mud No. 10" at 1.91 ft/sec, show that the "24" at the bottom refers to "0.00024 psi" (thus, the numbers in the plot, when multiplied by 10^{-5}, give the actual psi level). A high value of "83" is seen on the upper pipe surface; lows are generally obtained away from solid surfaces and at the annular floor. The average of these calculated values, taken over the bottom half of the annulus, supply the mean stress points on the horizontal axes of Figures 5-1, 5-2 and 5-3.

Becker *et al* (1989) noted that the best data fit, obtained through trial and error, was obtained with low shear rate parameters, in particular, Fann dial (stress) readings at low rotary speeds like 6 rpm. This corresponds to a shear rate of 10/sec. Our exact, computed results gave averaged rates of 7-9/sec for *all* the mud samples at 1.91 ft/sec; similarly, 11-14/sec at 2.86 ft/sec, and 14-19/sec at 3.82 ft/sec. Since these are in the 10/sec range, they explain why a 6 rpm correlation worked, at least in their particular test setup.

But in general, the Becker "low rpm" recommendation will *not* apply a priori; each nonlinear annular flow presents a unique physical problem with its own characteristic shears. In general, pipe to hole diameter ratio, as well as eccentricity, enter the equation. But this poses no difficulty since downhole properties *can* be obtained with minimal effort with the present program.

Cuttings removal in near-vertical holes with $\beta < 10^{\circ}$ is well understood; cleaning efficiency is proportional to annular velocity, or more precisely, the "Stokes product" between relative velocity and local viscosity. This product appeared naturally in Stokes' original low Reynolds flow solutions for flows past spheres, forming part of the coefficient describing net viscous drag.

For inclined wells, the usual notions regarding unimpeded settling velocities do not apply because different physical processes are at work. Cuttings travel almost immediately to the low side of the annulus, a consequence of gravity segregation; they remain there and form beds that may or may not slide downward. These truss or lattice-like structures have well defined mechanical yield stresses; the right amount of viscous friction will erode the cuttings bed, the same way mud circulation limits dynamic filter cake growth. This explains our success in using bottom-averaged viscous stress as the correlation parameter. The straight line fit also indicates that bed properties are linear in an elastic sense.

These ideas, of course, are not entirely new. Slurry pipeline designers, for example, routinely consider "boundary shear" and "critical tractive force." They have successfully modeled sediment beds as "series of superposed layers" with distinct yield strengths (Streeter, 1961). However, these studies are usually restricted to Newtonian carrier fluids in circular conduits.

While viscous shear emerges as the dominant transport parameter, its role was by no mean obvious at the outset. Other correlation quantities tested include vertical and lateral components of shear rates and stresses, axial velocity, apparent viscosity, and the Stokes product. These correlated somewhat well, particularly at low inclinations, but shear stress almost *always* worked. Take apparent viscosity, for example. Whereas Figure 6 of Becker *et al* (1989) shows significant wide-band scatter, listing *rotational viscometer* values ranging from 1 to 50 cp, our *exact* computations gave good correlations with *actual* apparent viscosities ranging up to 300 cp. Computed viscosities expectedly showed no meaningful connection to the apparent viscosities given by the University of Tulsa investigators, because the latter were inferred from unrealistic Fann dial readings. This point is illustrated quantitatively later.

We emphasize that Figures 5-1, 5-2, and 5-3 are based on unweighted muds. On a separate note, the effect of "pure changes in fluid density" should not alter computed shear stresses, at least theoretically, since the convective terms in the governing equations vanish for straight holes. In practice, however, oilfield weighting materials are likely to alter n and k; thus, some change in stress level might be anticipated. The effects of buoyancy, not treated here, will of course help without regard to changes in shear.

We have shown how cuttings concentration correlates in a satisfactory manner with *the mean viscous shear stress averaged over the lower half of the annulus*. Thus, impending hole-cleaning problems can be alleviated by first determining the existing average stress level, and then, adjusting n, k, and gpm values in the actual drilling fluid to increase that stress. Once this danger zone is past, additives can be used to reduce shear stress and hence mud pump pressure requirements. Simply increasing gpm may also help, although the effect of rheology on stress is probably more significant.

Interestingly, Seeberger *et al* (1989) described an important field study where extremely high velocities together with very high yield points did not alleviate hole cleaning problems. They suggested that extrapolated YP values may not be useful indicators of transport efficiency. Also, the authors pointed to the importance of elevated stress levels at low shear rates in cleaning large diameter holes at high angles. They experimentally showed how oil and water base muds having like rheograms, despite their obvious textural or "look and feel" differences, will clean with like efficiencies. This implies that a knowledge of n and k alone suffices in characterizing real muds.

The procedure suggested above requires minimal change to field operations. Standard viscometer readings, plotted on "log-log" paper or used in handbook formulas, still represent required information; but they should be used to determine actual downhole properties through computer analysis. Yield point and plastic viscosity, arising from older Bingham models, play no direct role in the present methodology although these parameters sometimes offer useful correlations. Also, results obtained from the eccentric model should be available in tabular form, and be accessible as software at the drilling site.

Discussion 2. Cuttings Transport Database

The viscometer properties and cuttings concentrations data for the 15 muds (at all angles and flow rates), together with exact computed results for shear rate, stress, apparent viscosity, annular speed, and Stokes product have been assembled into a comparative database for continuing study. These detailed results are available from the author upon request.

Tables 5-1, 5-2, and 5-3 summarize bottom-averaged results for the eccentric hole used in the Tulsa experiments. Computations show that the bottom of the hole supports a low shear rate flow, ranging from 10 to 20 reciprocal seconds. These values are consistent with the authors' low shear rate conclusions, established by trial and error from the experimental data. However, their rule of thumb is not universally correct; for example, the same muds and flow rates gave high shear rate results for several different downhole geometries.

Shear rates *can* vary substantially depending on eccentricity and diameter ratio. Direct computational analysis is the only legitimate and final arbiter. These tables also give calculated apparent viscosities along with values extrapolated from rotating viscometer data (shown in parentheses). Comparison shows that no correlation between the two exists, a result not unexpected, since the measurements bear little relation to the downhole flow. On the other hand, calculated apparent viscosities correlated well with cuttings concentration, although not as well as did viscous stress. This correlation was possible because bottom-averaged shear rates did not vary appreciably from mud to mud at any given flow speed. This effect may be fortuitous.

Table 5-1
Bottom-Averaged Fluid Properties @ 1.91 ft/sec

Mud	n	k lbf sec^n /in^2	Shear Rate 1/sec	Shear Stress (psi)	Apparent-Viscosity (cp)	
1	1.00	0.15E-6	9.1	0.13E-5	1	(1)
2	0.74	0.72E-5	8.1	0.29E-4	27	(8)
3	0.59	0.13E-4	7.8	0.34E-4	35	(5)
4	0.74	0.14E-4	8.3	0.59E-4	54	(15)
5	0.59	0.25E-4	7.6	0.67E-4	71	(9)
6	0.42	0.57E-4	7.4	0.95E-4	116	(6)
7	0.74	0.24E-4	8.1	0.97E-4	89	(25)
8	0.59	0.43E-4	7.6	0.11E-3	118	(15)
9	0.42	0.94E-4	7.5	0.16E-3	191	(10)
10	0.74	0.38E-4	8.2	0.16E-3	143	(40)
11	0.59	0.68E-4	7.7	0.18E-3	190	(24)
12	0.42	0.15E-3	7.5	0.25E-3	307	(16)
13	0.74	0.48E-4	8.0	0.19E-3	180	(50)
14	0.59	0.85E-4	7.6	0.22E-3	237	(30)
15	0.42	0.19E-3	7.4	0.32E-3	388	(20)

Table 5-2
Bottom-Averaged Fluid Properties @ 2.86 ft/sec

Mud	n	k lbf sec^n /in^2	Shear Rate 1/sec	Shear Stress (psi)	Apparent-Viscosity (cp)	
1	1.00	0.15E-6	14	0.20E-5	1	(1)
2	0.74	0.72E-5	12	0.39E-4	24	(8)
3	0.59	0.13E-4	11	0.42E-4	30	(5)
4	0.74	0.14E-4	12	0.78E-4	49	(15)
5	0.59	0.25E-4	11	0.84E-4	60	(9)
6	0.42	0.57E-4	11	0.11E-3	91	(6)
7	0.74	0.24E-4	12	0.13E-3	80	(25)
8	0.59	0.43E-4	11	0.14E-3	100	(15)
9	0.42	0.94E-4	11	0.19E-3	152	(10)
10	0.74	0.38E-4	12	0.21E-3	129	(40)
11	0.59	0.68E-4	11	0.23E-3	161	(24)
12	0.42	0.15E-3	11	0.30E-3	242	(16)
13	0.74	0.48E-4	12	0.26E-3	161	(50)
14	0.59	0.85E-4	11	0.28E-3	199	(30)
15	0.42	0.19E-3	11	0.38E-3	305	(20)

Table 5-3
Bottom-Averaged Fluid Properties @ 3.82 ft/sec

Mud	n	k lbf sec^n /in^2	Shear Rate 1/sec	Shear Stress (psi)	Apparent-Viscosity (cp)	
1	1.00	0.15E-6	18	0.27E-5	1	(1)
2	0.74	0.72E-5	16	0.49E-4	22	(8)
3	0.59	0.13E-4	15	0.50E-4	27	(5)
4	0.74	0.14E-4	17	0.98E-4	45	(15)
5	0.59	0.25E-4	15	0.10E-3	53	(9)
6	0.42	0.57E-4	15	0.13E-3	78	(6)
7	0.74	0.24E-4	16	0.16E-3	74	(25)
8	0.59	0.43E-4	15	0.17E-3	88	(15)
9	0.42	0.94E-4	15	0.21E-3	128	(10)
10	0.74	0.38E-4	16	0.26E-3	119	(40)
11	0.59	0.68E-4	15	0.27E-3	142	(24)
12	0.42	0.15E-3	15	0.34E-3	205	(16)
13	0.74	0.48E-4	17	0.33E-3	148	(50)
14	0.59	0.85E-4	15	0.33E-3	177	(30)
15	0.42	0.19E-3	15	0.43E-3	258	(20)

Discussion 3. Invert Emulsions Versus "All Oil" Muds

Recently, Conoco's Jolliet project successfully drilled a number of deviated wells, ranging 30° to 60° from vertical, in the deepwater Green Canyon Block 184 using a new "all oil" mud. Compared with wells previously drilled in the area with conventional invert emulsion fluids, the oil mud proved vastly superior with respect to cuttings transport and overall hole cleaning (Fraser, 1990a,b,c).

High levels of cleaning efficiency were maintained consistently throughout the drilling program. In this section we explain, using the fully predictive, eccentric annular flow model of Chapter 2, *why* the particular oil mud employed by Conoco performed well in comparison with the invert emulsion. The following discussion was first reported in Chin (1990c).

Given the success of the correlations developed in Discussion 2, it is natural to test our "stress hypothesis" under more realistic and difficult field conditions. Conoco's Green Canyon experience is ideal in this respect. Unlike the unweighted, bentonite-polymer, water-base muds used in the University of Tulsa experiments, the drilling fluids employed by Conoco were "invert emulsion" and "all oil" muds.

Again, Seeberger *et al.* (1989) have demonstrated how oil-base and water-base muds having like rheograms, despite obvious textural differences, will clean holes with like efficiencies. This experimental observation implies that a knowledge of n and k alone suffices in characterizing the carrying capacity of water, oil-base or emulsion-base drilling fluids. Thus, the use of a power law annular flow model as the basis for comparison for the two Conoco muds is completely warranted.

We assumed for simplicity a 2-inch radius drill pipe centered halfway down a 5-inch-radius borehole. This eccentricity is consistent with the 30° to 60° inclinations reported by Conoco. The n and k values we required were calculated from Figure 2 of Fraser (1990b), using Fann dial readings at 13 and 50 rpm. For the invert emulsion, we obtained n = 0.55 and k = 0.0001 lbf $\sec^n$/sq in; the values n = 0.21 and k = 0.00055 lbf $\sec^n$/sq in were found for the "all oil" mud.

Our annular geometry is identical to that used in Discussion 1 and in Becker *et al.* (1989). It was chosen so that the shear stress results obtained for the Tulsa water-base muds (shown in Figures 5-1 to 5-3) can be directly compared with those found for the weighted invert emulsion and oil fluids considered here.

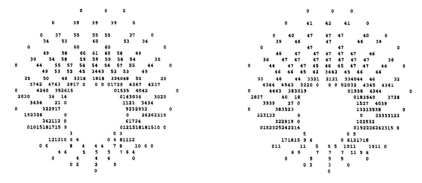

Figure 5-5a. Annular velocity, invert emulsion.

Figure 5-5b. Annular velocity, all oil mud.

For comparative purposes, the two runs described here were fixed at 500 gpm. To maintain this flow rate, the invert emulsion required a local axial pressure gradient of 0.010 psi/ft; Conoco's all oil mud, by contrast, required 0.029 psi/ft. Figures 5-5a and 5-5b, for invert emulsion and all oil muds, give calculated results for axial velocity in in/sec. Again, note how all no-slip conditions are identically satisfied.

Figures 5-6a and 5-6b display the absolute values of the *vertical* component of viscous shear stress; the leading significant digits are shown, corresponding to magnitudes that are typically $O(10^{-3})$ to $O(10^{-4})$ psi. This shear stress is obtained as the product of local apparent viscosity and shear rate, both of which vary throughout the cross-section. That is, the viscous stress is obtained *exactly* as "apparent viscosity $(x,y) \times dU(x,y)/dx$."

Figure 5-5a shows that the invert emulsion yields maximum velocities near 61 in/sec on the high side of the annulus; the maximums on the low side, approximately 5 in/sec, are less than ten times this value. By comparison, the "all oil" results in Figure 5-5b demonstrate how a smaller n tends to redistribute velocity more uniformly; still, the contrast is high, being 47 in/sec to 7 in/sec. The difference between the low side maximum velocities of 5 and 7 in/sec is not significant, and certainly does *not* explain observed large differences in cleaning efficiency.

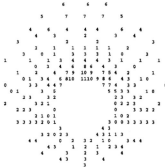
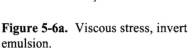

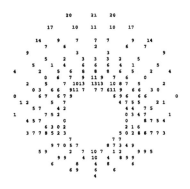

Figure 5-6a. Viscous stress, invert emulsion.

Figure 5-6b. Viscous stress, all oil mud.

Our earlier results in Discussion 1 provided experimental evidence suggesting that mean viscous shear stress is the correct correlation parameter for hole-cleaning efficiency. This is, importantly, again the case here. First note how Figure 5-6a gives a bottom radial stress distribution of "3-2-2-3-3" for the invert emulsion mud. In the case of Conoco's "all oil" mud, Figure 5-6b shows that these values significantly increase to "10-10-4-6-4."

We calculated mean shear stress values averaged over the lower half of the annulus. These values, for oil-base and invert-emulsion muds, respectively, were 0.00061 and 0.00027 psi. Their ratio, a sizable 2.3, substantiates the positive claims made in Fraser (1990b). Calculated shear stress averages for the University of Tulsa experiments in no case exceeded 0.0004 psi.

Similarly averaged apparent viscosities also correlated well, leading to a large ratio of 2.2 (the "apparent viscosities" in Becker *et al.* (1989) did not correlate at all, because unmeaningful rotational viscometer readings were used). Bottom-averaged shear rates, for oil-base and invert-emulsion muds, were calculated as 12.7 and 9.6/sec, respectively; at least in this case, we have again justified the "6 rpm (or 10/sec) recommendation" offered by many drilling practitioners. In general, however, shear rates will vary widely; they can be substantial depending on the particular geometry and drilling fluid.

The present results and the detailed findings of Discussion 1, together with the recommendations of Seeberger *et al* (1989), strongly suggest that "bottom-averaged" viscous shear stress correlates well with cuttings carrying capacity. Thus, as before, a driller suspecting cleaning problems should first determine his current downhole stress level; then he should alter n, k, and gpm to increase that stress. Once the danger is past, he can lower overall stress levels to reduce mud pump pressure requirements.

Discussion 4. Effect of Cuttings Bed Thickness

In vertical wells where drilled cuttings move unimpeded, cuttings transport and hole-cleaning efficiency vary directly as the product between mud viscosity and "relative particle and annular velocity." For inclined wells, bed formation introduces a new physical source for clogging. Often, this means that rules of thumb developed for vertical holes are not entirely applicable to deviated wells. For example, Seeberger *et al* (1989) pointed out that substantial increases in both yield point and annular velocity did not help in alleviating their hole problems. They suggested that high shear stresses at low shear rates would be desirable, and that stress could be a useful indicator of cleaning efficiency in deviated wells. We have given compelling evidence for this hypothesis.

Using the eccentric flow model of Chapter 2, we have demonstrated that "cuttings concentration" correlates *linearly* with mean shear stress, that is, *the viscous stress averaged over the lower half of the annulus*, for a wide range of oncoming flow speeds and well inclinations. Apparently, this empirical correlation holds for invert emulsions and oil-base muds as well.

Having established that shear stress is an important parameter in bed formation, it is natural to ask whether cuttings bed growth itself helps or hinders further growth; that is, does bed buildup constitute a self-sustaining, destabilizing process? The classic "ball on top of the hill," for instance, continually falls once it is displaced from its equilibrium position. In contrast, the "ball in the valley" consistently returns to its origin, demonstrating "absolute stability."

If cuttings bed growth itself induces further growth, the cleaning process would be unstable in the foregoing sense. This instability would underline, in field applications, the importance of controlling downhole rheology so as to increase stress levels at the onset of impending danger. Field site flow simulation could play an important role in operations, that is, in determining existing stress levels with a view towards optimizing fluid rheology in order to increase them. In this section, calculations are described that suggest that instability is possible.

In the eccentric flow calculations that follow, we assume a 2-inch-radius nonrotating drill pipe, displaced 1.5 inches downward in a 5-inch-radius borehole. This annular geometry is the same as the experimental setup reported in Becker *et al* (1989). For purposes of evaluation, we arbitrarily selected "Mud No. 10" used by the University of Tulsa team. It has a power law exponent of 0.736 and a consistency factor of 0.0000383 lbf sec^n/sq in. The total annular volume flow rate was fixed for all of our runs, corresponding to usual operating conditions. The average linear speed was held to 1.91 ft/sec or 22.9 in/sec. In the reported experiments, this speed yielded laminar flow at all inclination angles.

```
               0    0    0
            0   29   30   29    0
              0   28   44   45  44     28    0
             26   42        50      42   26
            0                50                  0
              39   48   51  51  51  48     39
            23  44  49   49  49  49  44     23
            0  34  45  46  44  44  44  46  45  34   0
              39  43  42  36  2636  42  43     39
            19   40   38  2514  1414  253138  40      19
             2933 3833  2213 0    0 0 01322 3338 3329
            0  3332  271912              01227  3233      0
            1523    28  10          0102328     2315
                2627    15 0         815   2726
            0    252213               6182225        0
              111720          0          0  20201711
                 1917  9 0             51319
              0 812141412 7        01012151412 8 0
                          2           0 2
                   1010 9 5 4        0 4 71010
                0 5      7    4   44   6 7      8 5 0
                   3 5      5    5    5   7 5 3
                0         4     4   4        0
                     0 2      2     0
                        0
```

Figure 5-7a. Annular velocity, "no bed."

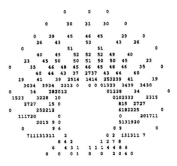

Figure 5-7c. Annular velocity, "medium bed."

```
                0    0    0
             0   30   31   30    0
               0   29   45   46  45     29    0
              26   43        52      43   26
             0              51              51          0
               40   49   52  52  52  49     40
              23   45  50      50  51  50  50  45     23
             0  35   46  48  45  46  46  45     35   0
               40  44  43  37  2737  43  44      40
             19   41   39  2514  1414  253239  41      19
              3034 3834  2313 0    0 0 01323 3438 3430
             0  34   282012              01228     34      0
             1523  3228  10          0102332     2315
                2627    15 0          815   2726
             0    252212               6172225          0
               111719          0         0   191711
                 191512  6             4121819
             0  710122210         0 6             0
                    6 5 1          0 2 8311210 7
                 6   4 2       1   6
                 5   4 3       3 2   4 7 6
                 0 0 3 0       3 3   4   5
                               0 1 0 0   0
```

Figure 5-7b. Annular velocity, "small bed."

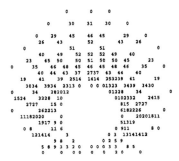

Figure 5-7d. Annular velocity, "large bed."

Four case studies were performed, the first containing no cuttings bed; then, assuming flat cuttings beds successively increasing in thickness. The level surfaces of the "small," "medium," and "large" beds were located at 0.4, 0.8, and 1.0 inch, respectively, from the bottom of the annulus. Required pressure drops varied from 0.0054 to 0.0055 psi/ft. As indicated in Chapter 2, the highly visual output format directly overlays computed quantities on the cross-sectional geometry, thus facilitating physical interpretation and correlation with annular position. Computed results for axial velocity in in/sec are shown above in Figures 5-7a to 5-7d. All four velocity distributions satisfy the no-slip condition exactly; the text plotter used, we note, does not always show 0's at solid boundaries because of character spacing issues. The "No Bed" flow given in Figure 5-7a demonstrates very clearly how velocity can vary rapidly about the annulus. For example, it has maximums of 51 and 5 in/sec above and below the pipe, a ten-fold difference. Figures 5-7b to 5-7d show that this factor increases, that is, worsens, as the cuttings bed increases in thickness.

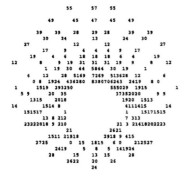

Figure 5-8a. Viscous stress, "no bed."

Figure 5-8c. Viscous stress, "medium bed."

Figure 5-8b. Viscous stress, "small bed."

Figure 5-8d. Viscous stress, "large bed."

Figures 5-8a to 5-8d give computed results for the vertical component of the shear stress, that is, "apparent viscosity (x,y) × strain rate dU/dx," where x increases downward. Results for the stress related to "dU/dy," not shown because of space limitations, behaved similarly. For clarity, only the absolute values are displayed; the actual values, which are separately available in tabulated form, vary from $O(10^{-4})$ to $O(10^{-3})$ psi.

Note how the bottom viscous stresses decrease in magnitude as the cuttings bed builds in thickness. This decrease, which is accompanied by decreases in throughput area, further compounds cuttings transport problems and decreases cleaning efficiency. Thus, hole clogging is a self-sustaining, destabilizing process. Unless the mud rheology itself is changed in the direction of increasing stress, differential sticking and stuck pipe are possible. This decrease of viscous stress with increasing bed thickness is also supported experimentally. Quigley *et al.* (1990) measured "unexpected" decreases in fluid (as opposed to mechanical) friction in a carefully controlled flow loop where cuttings beds were allowed to grow. While concluding that "cuttings beds can reduce friction," the authors clearly do not recommend its application in the field, as it increases the possibility of differential sticking.

Numerical results such as those shown in Figures 5-8a to 5-8d provide a quantitative means for comparing cleaning capabilities between different muds at different flow rates. "Should I use the 'high tech' mud offered by Company A when the simpler drilling fluid of Company B, run at a different speed, will suffice?" With numerical simulation, these and related questions are readily answered. The present results indicate that the smaller the throughput height, the smaller the viscous stresses will be.

This is intuitively clear since narrow gaps impose limits upon the peak bottom velocity and hence the maximum stress. We caution that this result applies only to the present calculations and may not hold in general. The physical importance of cuttings beds indicates that they should be modeled in any serious well planning activity. This necessity also limits the potential of recently developed bipolar coordinate annular flow models. These handle circular eccentric annular geometries well, but they cannot be generalized to handle more difficult holes with cuttings beds.

Discussion 5. Why 45⁰ - 60⁰ Inclinations Are Worst

Various experimenters, e.g., Becker *et al.* (1989) and Brown *et al.* (1989), have reported especially severe hole-cleaning problems for deviated wells inclined approximately 45° to 60° from the vertical. The experimental data reported in the former paper, shown in our Figures 5-1, 5-2 and 5-3, indicate that cuttings concentration for a given flow speed peaks somewhere between $\beta = 30^\circ$ and 70°, where β is measured from the vertical. These measurements appear to be reliable and repeatable, and similar results have been reproduced at a number of independent test facilities.

One might ask why the expected worsening with increased inclination angle β should not vary monotonically. Why should cuttings concentration at first increase, and then decrease? This relative maximum is easily understood by noting that the *net* cuttings concentration C for a prescribed flow rate at a given angle must depend on both vertical and horizontal hole-cleaning mechanisms. In fact, C should be weighted by resolving it into component contributions parallel and orthogonal to the well axis, taking a vertical value C_v at $\beta = 0^\circ$ and a horizontal value C_h at $\beta = 90^\circ$. Thus, we have

$$C = C_h \sin \beta + C_v \cos \beta > 0 \qquad (5\text{-}1)$$

To obtain relative maxima and minima, the usual rules of calculus require us to differentiate Equation 5-1 with respect to β, and set the result to zero. That is, set

$$dC/d\beta = C_h \cos \beta - C_v \sin \beta = 0 \qquad (5\text{-}2)$$

to obtain the critical angle and concentration

$$\beta_{cr} = \arctan C_h/C_v \qquad (5\text{-}3)$$

$$C_{max} = C_h/\sin \beta_{cr} \qquad (5\text{-}4)$$

That the critical concentration is a maximum is easily seen from the fact that

$$d^2C/d\beta^2 = - C_h \sin \beta - C_v \cos \beta = - C < 0 \qquad (5\text{-}5)$$

is negative. Let us apply the data obtained of Becker *et al* (1989). From Figure 5-3 previously, we estimate C_h = 25%; from Figure 5-1, we take C_v = 12%. Substitution in Equation 5-3 yields β_{cr} = 64° and a corresponding value of C_{max} = 28%. The calculated 64° agrees with observation, while the 28% concentration is consistent with the high 45° concentration results in Figure 5-2.

In general, once C_h and C_v are individually known from horizontal and vertical flow loop tests, it is possible using Equations 5-3 and 5-4 to determine the worst case inclination β_{cr} and its C_{max} for that particular mud and flow rate. In practice there may be some slight dependence of C_h on β, since bed yield stresses may depend on gravity orientation and sedimentary packing. Note that β is related to the α of Chapters 2, 3 and 4 by $\alpha + \beta = 90^o$. We emphasize that our "worst case" analysis applies only to unidirectional flows. When gravity segregation is important, the annular model developed in Chapter 4 may be more pertinent; the reverse flows possible for certain channel parameters only worsen as the hole grows more horizontal.

Discussion 6. Key Issues in Cuttings Transport

The empirical cuttings transport literature contains confusing observations and recommendations that, in light of the foregoing results and those of Chapter 3, can be easily resolved. The references at the end of the chapter provide a cross-section of recent experimental results and industry views, although our list is by no means exhaustive or comprehensive. We will address several questions commonly raised by drillers.

First and foremost is, "Which parameters control transport efficiency?" In vertical wells, the drag or uplift force on small isolated chips can be obtained from lubrication theory via Stoke's or Oseen's low Reynolds number equations. This force is proportional to the product between local viscosity and the first power of relative velocity between chip and fluid. The so-called "Stokes product" correlates well in vertical holes.

In deviated and horizontal holes with eccentric annular geometries, cuttings beds invariably form on the low side. These beds consist of well-defined mechanical structures with nonzero yield stresses; to remove or erode them, viscous fluid stresses must be sufficiently strong to overcome their resilience. The stresses computed on a laminar basis are sufficient for practical purposes, because low side, low velocity flows are almost always laminar. In this sense, any turbulence in the high side flow is unimportant since it plays no direct role in bed removal (the high side flow does convect debris that are uplifted by rotation). This observation is reiterated by Fraser (1990c). In his paper, Fraser correctly points out that too much significance is often attached to velocity criteria and fluid turbulence in deviated wells.

A second common question concerns the role of drillpipe rotation. With rotation, centrifugal effects throw cuttings circumferentially upwards where they are convected uphole by the high side flow; then they fall downwards. In the first part of this cycle, the cuttings are subject to drag forces not unlike those found in vertical wells. Here turbulence can be important, determining the amount of axial throw traversed before the cuttings are redeposited into the bed. Order of magnitude estimates comparing rotational to axial effects can be obtained using the formulas in Chapter 3.

Other effects of rotation are subtly tied to the rheology of the background fluid. Conflicting observations and recommendations are often made regarding drillpipe rotation for concentric annuli. To resolve them, we need to reiterate some theoretical results of Chapter 3. There we demonstrated that axial and circumferential speeds completely decouple for laminar Newtonian flows despite the nonlinearity of the Navier-Stokes equations. This is so because the convective terms exactly vanish, allowing us to "naively" (but correctly) superpose the two orthogonal velocity fields.

This fact was, apparently, first deduced by Savins *et al.* (1966), who noted that no coupling between the discharge rate, axial pressure gradient, relative rotation, and torque could be found through the viscosity coefficient for Newtonian flows. This author is indebted to J. Savins for directing him to the earlier literature. The decoupling implies that experimental findings obtained using Newtonian drilling fluids (primarily water and air) cannot be extrapolated to more general power law or Bingham plastic rheologies. Likewise, rules of thumb deduced using real drilling muds will not be consistent with those found for water. Newtonian (e.g., brines) and "real" muds behave differently in the presence of pipe rotation. In a Newtonian fluid, rotation will not affect the axial flow, although centrifugal "throwing" is still important.

In an initially steady non-Newtonian flow where the mudpump is operating at constant pressure, a momentary increase in rpm leads to a temporary surge in flow rate and thus improved hole cleaning. But once the pump readjusts itself to the prescribed gpm, this advantage is lost unless, obviously, the discharge rate itself is reset upwards or the rheology is improved by using a mud additive.

The decoupling discussed above applies to Newtonian flows in concentric annuli only. The coupling between axial and circumferential velocities reappears, even for Newtonian flows, when the rotating motion occurs in an eccentric annulus. This is so because the nonlinear convective terms will not identically vanish. This isolated singularity suggests that concentric flow loop tests using Newtonian fluids provide little benefit or information in terms of field usefulness. In fact, their results will be subject to misinterpretation.

And the role of fluid rheology? We have demonstrated how bottom-averaged shear stress can be used as a meaningful correlation parameter for cuttings transport in eccentric deviated holes. This mean viscous stress can be computed using the method developed in Chapter 2. The arguments given in Discussions 1, 2, and 3 are sound on physical grounds; in cuttings transport, rheology is a significant player by way of its effect on fluid stress.

Note that we have *not* modeled the dynamics of single chips or ensembles of cuttings. Nor are such analyses recommended; for field applications, it is only necessary to use stress as a correlation parameter. Modeling the dynamics of aggregates of chips involves mathematics so complicated that it is difficult to anticipate any practical significance, even in the long term. Finally, a comment on the role of increased fluid density in improving hole cleaning. This is undeniably the rule, since higher densities increase buoyancy effects; it applies to all flows, whether or not they are annular or deviated.

EVALUATION OF SPOTTING FLUIDS FOR STUCK PIPE

Stuck pipe due to differential pressure between the mud column and the formation often results in costly time delays. The mechanics governing differential sticking are well known (Outmans, 1958). In the past, diesel oil, mineral oil, and mixtures of these with surfactants, clays, and asphalts were usually spotted to facilitate the release of the drill string. However, the use of these conventional spotting fluids is now stringently controlled by government regulation; environmentally safe alternatives must be found.

Recently, Halliday and Clapper (1989) described the development of a successful, non-toxic, water-base system. Their new spotting fluid, identified using simple laboratory screening procedures, was used to free a thousand feet of stuck pipe in a 39° hole, from a sand section in the Gulf of Mexico. Since water-base spotting fluids, being relatively new, have seldom been studied in the literature, it is natural to ask whether or not they really work; and, if so, how. This section calculates, on an exact eccentric flow basis, three important mechanical properties, namely, the apparent viscosity, shear stress and shear rate of the drilling mud, with and without the spot additive. Then we provide a complete physical explanation for the reported success. The spotting fluid essentially works by mechanically reducing overall apparent viscosity; this enables the resultant fluid to perform its chemical functions better. The results of this section were first reported in Chin (1991).

The eccentric borehole annular flow model of Chapter 2 was used. While we have successfully applied it to hole cleaning before the occurrence of stuck pipe, it is of interest to apply it to other drilling problems, for example, determining the effectiveness of spotting fluids in freeing stuck pipe. Which mechanical properties are relevant to spotted fluids? What should their orders of magnitude be? We examined the water-base system described in Halliday and Clapper (1989) because such systems are becoming increasingly important. Why they work is not yet thoroughly understood. But it suffices to explain how the water-base spotting fluid behaves, insofar as mechanical fluid properties are concerned, on a single-phase, miscible flow basis. Conventional capillary pressure and multiphase considerations for "oil on aqueous filter cake" effects do not apply here, since we are dealing with "water on water" flows.

We performed our calculations for a 7.75-inch-diameter drill collar located eccentrically within a 12.5-inch-diameter borehole. This corresponds to the bottomhole assembly reported by the authors. A small bottom annular clearance of 0.25 inches was selected for evaluation purposes. This almost closed gap is consistent with the impending stuck pipe conditions characteristic of typical deviated holes. The authors' Table 11 gives Fann 600 and 300 rpm dial readings for the water-base mud used, before and after spot addition; both fluids, incidentally, were equal in density. In the former case, these values were 46 and 28; in the latter, 41 and 24. These properties were measured at 120° F. The calculated n and k power law coefficients are, respectively, 0.70 and 0.000025 lbf $\sec^n$/sq in for the original mud; for the spotted mud, we obtained 0.77 and 0.0000137 lbf $\sec^n$/sq in.

Halliday and Clapper reported that attempts to free the pipe by jarring down, with the original drilling fluid in place, were unsuccessful. At that point, the decision to spot the experimental non-oil fluid was made. Since jarring operations are more impulsive, rather than constant pressure drop processes, we calculated our flow properties for a wide range of applied pressure gradients. Note that the unsteady, convective term in the governing momentum equation has the same physical dimensions as pressure gradient. It was in this approximate engineering sense that our exact simulator was used.

The highest pressure gradients shown below correspond to volume flow rates near 1,100 gpm. Computed results for several parameters averaged over the lower half of the annulus are shown in Tables 5-4 and 5-5.

Table 5-4
Fluid Properties, Original Mud

Pressure Gradient (psi/ft)	Flow Rate (gpm)	Apparent- Viscosity (lbf sec/in^2)	Shear Rate (sec^{-1})	Viscous Stress (psi)
0.0010	69	0.000036	0.4	0.000011
0.0020	185	0.000027	1.2	0.000022
0.0030	329	0.000022	2.1	0.000033
0.0035	410	0.000021	2.6	0.000038
0.0040	497	0.000020	3.2	0.000044
0.0050	683	0.000018	4.3	0.000055
0.0060	886	0.000017	5.6	0.000066
0.0070	1105	0.000016	7.0	0.000077

Table 5-5
Fluid Properties, Spotted Mud

Pressure Gradient (psi/ft)	Flow Rate (gpm)	Apparent- Viscosity (lbf sec/in^2)	Shear Rate (sec^{-1})	Viscous Stress (psi)
0.0010	140	0.000014	1.0	0.000011
0.0020	344	0.000012	2.4	0.000022
0.0023	412	0.000011	2.8	0.000025
0.0030	582	0.000010	4.0	0.000033
0.0035	711	0.000010	4.9	0.000039
0.0040	846	0.000010	5.8	0.000044
0.0050	1130	0.000009	7.8	0.000055

We emphasize that calculated averages are sensitive to annular geometry; thus, the results shown in Tables 5-4 and 5-5 may not apply to other borehole configurations. In general, any required numerical quantities should be recomputed with the exact downhole geometry.

The results for averaged shear stress are "almost" Newtonian in the sense that stress increases linearly with applied pressure gradient. This unexpected outcome is not generally true of non-Newtonian flows. Both treated and untreated muds, in fact, show exactly the same shear stress values. However, shear rate and volume flow rate results for the two muds vary differently, and certainly nonlinearly with pressure gradient. The most interesting results, those concerned with spotting properties, are related to apparent viscosity.

The foregoing calculations importantly show how the apparent viscosity for the spotted mud, which varies spatially over the annular cross-section, has a nearly constant "bottom average" near 0.000010 lbf sec/in^2 over the entire range of flow rates. This value is approximately 69 cp, far in excess of the viscosities inferred from rotational viscometer readings, but still *two to three times less* than those of the original untreated mud. The importance of "low viscosity" in spotting fluids is emphasized in several mud company publications brought to this author's attention. Whether the apparent viscosity is high or low, of course, cannot be determined independently of the hole geometry and the prescribed pressure gradient.

The apparent viscosity is relevant because it is related to the lubricity factor conventionally used to evaluate spotting fluids. It is importantly calculated on a true eccentric flow basis, rather than determined from (unrelated) rotational viscometer measurements. As in cuttings transport, viscometer measurements are only valid to the extent that they provide accurate information for determining n and k over a limited range of shear rates.

That the treated fluid exhibits much lower viscosities over a range of applied pressures is consistent with its ability to penetrate the pipe and mudcake interface. This lubricates and separates the contact surfaces over a several-hour period; thus, it enables the spotting to perform its chemical functions efficiently, thereby freeing the stuck drill string. The effectiveness of any spotting fluid, of course, must be determined on a case by case basis.

While computed averages for apparent viscosity are almost constant over a range of pressure gradients, we emphasize that exact cross-sectional values for each flow property can be quite variable. For example, consider the annular flow for the spotted mud under a pressure gradient of 0.002 psi/ft, with a corresponding flow rate of 344 gpm. The velocity solutions in in/sec, using the highly visual output format discussed in Chapter 2, are shown in Figure 5-9; note, again, how no-slip conditions are rigorously enforced at all solid surfaces.

Figure 5-9. Annular velocity.

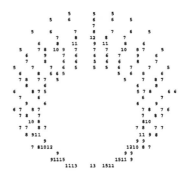

Figure 5-10. Apparent viscosity.

Figure 5-10 gives results for *exact apparent viscosity*, which varies with spatial position, again plotted over the highly eccentric geometry itself. Although the text plotter used does not furnish sufficient visual resolution at the bottom of the annular gap, reference to tabulated solutions indicates pipe surface values of "13," increasing to "29" at the midsection, finally decreasing to 13×10^{-6} lbf sec/in^2 at the borehole wall. The flatness of the cuttings bed, or the extent to which it modifies annular bottom geometry, will also be an important factor as far as lubricity is concerned. Any field-oriented hydraulics simulation should also account for such bed effects.

We had demonstrated earlier that flow modeling can be used to correlate laboratory and field cuttings transport efficiency data against actual (computed) downhole flow properties. Bottom-averaged viscous shear stress importantly emerged as the physically significant correlation parameter. The present section indicates that annular flow modeling can also be used to evaluate the effectiveness of spotting fluids in freeing stuck pipe. Here, the important correlation parameter is average apparent viscosity, a fact most mechanical engineers might have anticipated. This quantity is directly related to the lubricity factor usually obtained in laboratory measurements.

CEMENTING APPLICATIONS

The modeling of annular borehole flows for drilling applications is similar to that for cementing. Aside from obvious differences associated with "touch and feel" contrasts between drilling muds and cement slurries, little in the way of analysis changes. What differs, however, lies in the way computed quantities are used. In drilling problems, "mean viscous shear stress" and average velocity control cuttings transport efficiency, depending on the deviation of the well. On the other hand, a low value of "mean apparent viscosity" appears to determine the effectiveness of a spotted mud for use in releasing stuck pipe.

Typical references provided later furnish a representative cross-section of the modern cementing literature. The primary operational concern is effective mud displacement and removal. The industry presently emphasizes the importance of good rheology and high velocity, but these qualities alone are not sufficient. In order to produce good displacements, the stability of the cement velocity profile with respect to disturbances induced by the upstream mud should be addressed.

The cement velocity profile must be hydrodynamically stable and robust. Unstable velocity distributions may break down rapidly into viscous fingers and channel prematurely. An analogous problem in reservoir engineering is found in waterflooding: the displacement front may disperse into tiny fingers that propagate into the downstream flow when adverse mobility ratios are encountered. There the problem is solved by using flow additives whose attributes are determined by detailed mathematical modeling.

The classic monograph of Lin (1967) explains how the stability characteristics of any particular flow can be obtained as solutions to the so-called Rayleigh or Orr-Sommerfeld equations. Good velocity profiles in this sense can be ascertained by coupling the work of Chapter 2, which generates velocity profiles, to stability models that evaluate their ability to withstand disturbances. Stability analyses are routinely used in aeronautical and chemical engineering, for example, in the study of turbulent transition on wing surfaces and in ducts. They are also important in different secondary recovery aspects of reservoir engineering. More research should be directed towards this area. In the remainder of this section, typical velocity profiles are generated for comparative purposes only using the models of Chapters 2 and 3. These are not evaluated with respect to hydrodynamic stability.

Example 1. Eccentric Nonrotating Flow, Baseline Concentric Case

---

We first establish a simple concentric solution as the basis for further annular flow comparison. We will assume for the casing outer diameter a "pipe radius" of 3.0 in and a borehole radius of 4.5 in. We will evaluate the behavior of two cement slurries. The first is an API Class H slurry with power law coefficients $n = 0.30$ and $k = 0.001354$ lbf sec^n/in^2, while the second is a Class C slurry having $n = 0.43$ and $k = 0.0002083$ lbf sec^n/in^2. We shall refer to these as our "Class H" and "Class C" flows.

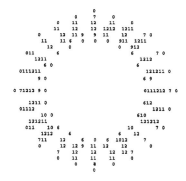

Figure 5-11a. Annular velocity, Class H slurry.

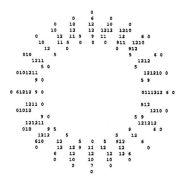

Figure 5-11c. Annular velocity, Class C slurry.

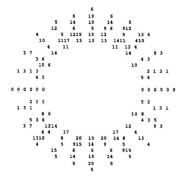

Figure 5-11b. Viscous stress, Class H slurry.

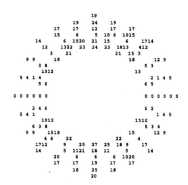

Figure 5-11d. Viscous stress, Class C slurry.

For convenience only, we will fix in our comparisons throughout, an approximate value for the pressure gradient, say dP/dz = - 0.1 psi/ft, arbitrarily chosen. Thus, some predicted gpms may be excessive from an engineering standpoint. The simulations described in Examples 1-4 below are run "as is" in the comparative sense of Chapter 2, and no attempt has been made to "fine tune" or calibrate the variable mesh to any known solution. The computed velocity distribution in in/sec for the Class H slurry is shown in Figure 5-11a. The corresponding volume flow rate is 8.932 gpm. The stress formed by the product "apparent viscosity (x,y) $\times$ dU(x,y)/dx" appears in Figure 5-11b. Typically, these stresses might be 0.5×10^{-3} psi in magnitude. A run summary is given in Table 5-6.

In the next example, we will rerun the simulation for the Class C slurry. The volume flow rate obtained in this case is 849.1 gal/min. The computed velocity and stress profiles are shown in Figure 5-11c (where the "12" indicates 120 in/sec) and Figure 5-11d. Run summaries for averaged quantities are given in Table 5-7.

Table 5-6
Example 1: Summary, Average Quantities (Class H Slurry)

```
TABULATION OF CALCULATED AVERAGE QUANTITIES:
Area weighted means of absolute values taken over
BOTTOM HALF of annular cross-section ...
o  Average annular velocity = .8955E+00 in/sec
o  Average apparent viscosity = .2378E-02 lbf sec/sq in
o  Average stress, AppVis x dU/dx, = .8820E-03 psi
o  Average stress, AppVis x dU/dy, = .7530E-03 psi
o  Average dissipation = .3112E-02 lbf/(sec sq in)
o  Average shear rate dU/dx = .1500E+01 1/sec
o  Average shear rate dU/dy = .1278E+01 1/sec
o  Average Stokes product = .2908E-02 lbf/in

TABULATION OF CALCULATED AVERAGE QUANTITIES, II:
Area weighted means of absolute values taken over
ENTIRE annular (x,y) cross-section ...
o  Average annular velocity = .8937E+00 in/sec
o  Average apparent viscosity = .2397E-02 lbf sec/sq in
o  Average stress, AppVis x dU/dx, = .8030E-03 psi
o  Average stress, AppVis x dU/dy, = .7964E-03 psi
o  Average dissipation =' .3076E-02 lbf/(sec sq in)
o  Average shear rate dU/dx = .1362E+01 1/sec
o  Average shear rate dU/dy = .1349E+01 1/sec
o  Average Stokes product = .2925E-02 lbf/in
```

Table 5-7

Example 1: Summary, Average Quantities (Class C Slurry)

```
TABULATION OF CALCULATED AVERAGE QUANTITIES:
Area weighted means of absolute values taken over
BOTTOM HALF of annular cross-section ...
O  Average annular velocity = .8511E+02 in/sec
O  Average apparent viscosity = .1816E-04 lbf sec/sq in
O  Average stress, AppVis x dU/dx, = .1190E-02 psi
O  Average stress, AppVis x dU/dy, = .1015E-02 psi
O  Average dissipation = .4555E+00 lbf/(sec sq in)
O  Average shear rate dU/dx = .1436E+03 1/sec
O  Average shear rate dU/dy = .1223E+03 1/sec
O  Average Stokes product = .2027E-02 lbf/in

TABULATION OF CALCULATED AVERAGE QUANTITIES, II:
Area weighted means of absolute values taken over
ENTIRE annular (x,y) cross-section ...
O  Average annular velocity = .8497E+02 in/sec
O  Average apparent viscosity = .1826E-04 lbf sec/sq in
O  Average stress, AppVis x dU/dx, = .1082E-02 psi
O  Average stress, AppVis x dU/dy, = .1073E-02 psi
O  Average dissipation = .4502E+00 lbf/(sec sq in)
O  Average shear rate dU/dx = .1304E+03 1/sec
O  Average shear rate dU/dy = .1291E+03 1/sec
O  Average Stokes product = .2036E-02 lbf/in
```

Example 2. Eccentric Nonrotating Flow, Eccentric Circular Case

The same pipe and hole sizes used in Example 1 are assumed here, but the casing is displaced downward, resting off the bottom by 0.5 in. Both the casing and the borehole are perfect circles. For the Class H slurry, a flow rate of 48.4 gpm is obtained, exceeding the 8.93 gpm of Example 1. Self-explanatory computed results are given for comparative purposes. In Figure 5-12a the "9" indicates 9 in/sec, while in Figure 5-12b the stresses are typically 0.001 psi. Averaged results appear in Table 5-8.

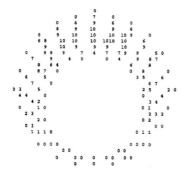

Figure 5-12a. Annular velocity, Class H slurry.

Figure 5-12b Viscous stress, Class H slurry.

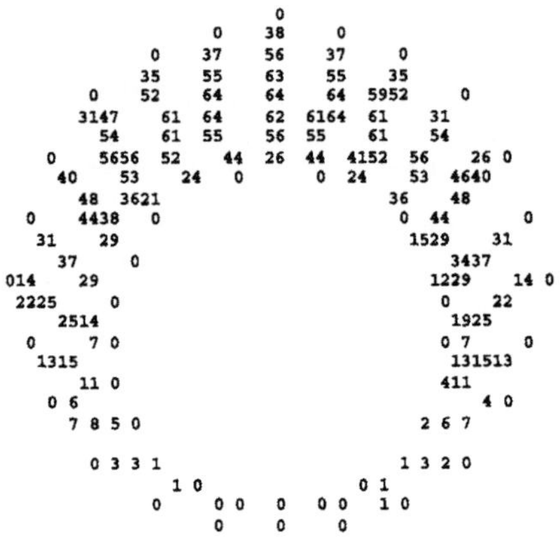

```
                        0
                  0    38    0
             0   37   56   37    0
            35   55   63   55   35
        0   52   64   64   64  5952    0
     3147   61   64   62  6164   61   31
       54   61   55   56   55   61   54
   0  5656   52   44   26   44  4152   56   26 0
     40   53   24    0        0   24   53 4640
     48  3621                      36   48
   0  4438    0                  0  44         0
    31   29                          1529   31
     37    0                              3437
   014   29                           1229   14 0
   2225    0                          0   22
     2514                               1925
   0    7 0                           0 7        0
    1315                               131513
       11 0                            411
   0  6                                     4 0
     7 8 5 0                          2 6 7

        0 3 3 1                    1 3 2 0
             1 0                  0 1
           0     0 0    0    0 0   1 0
                  0    0    0
```

Figure 5-12c. Annular velocity, Class C slurry.

Table 5-8
Example 2: Summary, Average Quantities (Class H Slurry)

```
TABULATION OF CALCULATED AVERAGE QUANTITIES:
Area weighted means of absolute values taken over
BOTTOM HALF of annular cross-section ...
O  Average annular velocity = .5958E+00 in/sec
O  Average apparent viscosity = .2206E-02 lbf sec/sq in
O  Average stress, AppVis x dU/dx, = .8796E-03 psi
O  Average stress, AppVis x dU/dy, = .1014E-02 psi
O  Average dissipation = .4405E-02 lbf/(sec sq in)
O  Average shear rate dU/dx = .1097E+01 1/sec
O  Average shear rate dU/dy = .1796E+01 1/sec
O  Average Stokes product = .5406E-03 lbf/in

TABULATION OF CALCULATED AVERAGE QUANTITIES, II:
Area weighted means of absolute values taken over
ENTIRE annular (x,y) cross-section ...
O  Average annular velocity = .3470E+01 in/sec
O  Average apparent viscosity = .1302E-02 lbf sec/sq in
O  Average stress, AppVis x dU/dx, = .1171E-02 psi
O  Average stress, AppVis x dU/dy, = .1201E-02 psi
O  Average dissipation = .1560E-01 lbf/(sec sq in)
O  Average shear rate dU/dx = .4129E+01 1/sec
O  Average shear rate dU/dy = .4283E+01 1/sec
O  Average Stokes product = .2691E-02 lbf/in
```

The calculations for this circular eccentric casing and hole are now repeated for the Class C slurry. Here, the volume flow rate is 2,706 gal/min. Computed results for the velocity distribution are shown in Figure 5-12c, where "60" indicates 600 in/sec; averaged results are recorded in Table 5-9.

Table 5-9
Example 2: Summary, Average Quantities (Class C Slurry)

```
TABULATION OF CALCULATED AVERAGE QUANTITIES:
Area weighted means of absolute values taken over
BOTTOM HALF of annular cross-section ...
O  Average annular velocity = .3340E+02 in/sec
O  Average apparent viscosity = .2407E-04 lbf sec/sq in
O  Average stress, AppVis x dU/dx, = .8455E-03 psi
O  Average stress, AppVis x dU/dy, = .9885E-03 psi
O  Average dissipation = .2660E+00 lbf/(sec sq in)
O  Average shear rate dU/dx = .6535E+02 1/sec
O  Average shear rate dU/dy = .9999E+02 1/sec
O  Average Stokes product = .5218E-03 lbf/in

TABULATION OF CALCULATED AVERAGE QUANTITIES, II:
Area weighted means of absolute values taken over
ENTIRE annular (x,y) cross-section ...
O  Average annular velocity = .1930E+03 in/sec
O  Average apparent viscosity = .1542E-04 lbf sec/sq in
O  Average stress, AppVis x dU/dx, = .1366E-02 psi
O  Average stress, AppVis x dU/dy, = .1369E-02 psi
O  Average dissipation = .1176E+01 lbf/(sec sq in)
O  Average shear rate dU/dx = .2361E+03 1/sec
O  Average shear rate dU/dy = .2333E+03 1/sec
O  Average Stokes product = .2112E-02 lbf/in
```

Example 3. Eccentric Nonrotating Flow, A Severe Washout

Here the eccentric geometry of Example 2 is modified by including a severe "washout" at the top of the hole. Washouts are typically caused by gravity loosening of unconsolidated sands in the presence of high velocity fluid motion. The washout is nonsymmetrical; it may, for example, have resulted from the rotating action of the drillbit against an unconsolidated sand. We could just as easily have modeled a keyseat indentation or any other wall deformation, of course. Average properties for the Class H and Class C flows are listed in Tables 5-10 and 5-11, respectively.

The flow rate for the Class H slurry is 274.9 gpm, exceeding the 48.4 gpm of Example 2. The corresponding velocity results are displayed in Figure 5-13a, where "50" indicates 50 in/sec. The simulations for the Class C slurry gave a high volume flow rate of 8,819 gpm; detailed velocity distributions are shown in Figure 5-13b where "20" means 2,000 in/sec. Note that Figures 13a,b are presented together with Figures 14a,b,c,d in order to conserve space.

Table 5-10
Example 3: Summary, Average Quantities (Class H Slurry)

```
TABULATION OF CALCULATED AVERAGE QUANTITIES:
Area weighted means of absolute values taken over
BOTTOM HALF of annular cross-section ...
O  Average annular velocity = .8670E+00 in/sec
O  Average apparent viscosity = .1982E-02 lbf sec/sq in
O  Average stress, AppVis x dU/dx, = .9911E-03 psi
O  Average stress, AppVis x dU/dy, = .1123E-02 psi
O  Average dissipation = .7581E-02 lbf/(sec sq in)
O  Average shear rate dU/dx = .1636E+01 1/sec
O  Average shear rate dU/dy = .2620E+01 1/sec
O  Average Stokes product = .5448E-03 lbf/in

TABULATION OF CALCULATED AVERAGE QUANTITIES, II:
Area weighted means of absolute values taken over
ENTIRE annular (x,y) cross-section ...
O  Average annular velocity = .1338E+02 in/sec
O  Average apparent viscosity = .9957E-03 lbf sec/sq in
O  Average stress, AppVis x dU/dx, = .1683E-02 psi
O  Average stress, AppVis x dU/dy, = .1720E-02 psi
O  Average dissipation = .8241E-01 lbf/(sec sq in)
O  Average shear rate dU/dx = .1402E+02 1/sec
O  Average shear rate dU/dy = .1285E+02 1/sec
O  Average Stokes product = .2727E-02 lbf/in
```

Table 5-11
Example 3: Summary, Average Quantities (Class C Slurry)

```
TABULATION OF CALCULATED AVERAGE QUANTITIES:
Area weighted means of absolute values taken over
BOTTOM HALF of annular cross-section ...
O  Average annular velocity = .3685E+02 in/sec
O  Average apparent viscosity = .2351E-04 lbf sec/sq in
O  Average stress, AppVis x dU/dx, = .8819E-03 psi
O  Average stress, AppVis x dU/dy, = .1025E-02 psi
O  Average dissipation = .3129E+00 lbf/(sec sq in)
O  Average shear rate dU/dx = .7239E+02 1/sec
O  Average shear rate dU/dy = .1105E+03 1/sec
O  Average Stokes product = .5244E-03 lbf/in

TABULATION OF CALCULATED AVERAGE QUANTITIES, II:
Area weighted means of absolute values taken over
ENTIRE annular (x,y) cross-section ...
O  Average annular velocity = .4319E+03 in/sec
O  Average apparent viscosity = .1338E-04 lbf sec/sq in
O  Average stress, AppVis x dU/dx, = .1718E-02 psi
O  Average stress, AppVis x dU/dy, = .1781E-02 psi
O  Average dissipation = .3035E+01 lbf/(sec sq in)
O  Average shear rate dU/dx = .4546E+03 1/sec
O  Average shear rate dU/dy = .4245E+03 1/sec
```

Example 4. Eccentric Nonrotating Flow, Casing with Centralizers.

Centralizers are typically used to prevent excessively low velocity cement zones from forming; the annular flow is concentric, more or less, and eccentricity is avoided at all costs. In this final example, we reconsider the concentric annular borehole flow of Example 1; however, we will introduce four centralizers. The geometry is essentially the same for drill collars with stabilizers. Again, the pipe and borehole radii are, respectively, 3.0 in and 4.5 in. For clarity, the screen displays for both the annular geometry and its corresponding computational mesh are duplicated in Figures 5-14a and 5-14b. Note how our centralizers are modeled by indenting the borehole contour inward. Alternatively, we could have deformed the pipe contour outward.

For the Class H slurry, the computed flow rate for this configuration is 8.30 gpm, which is slightly less than the value obtained without centralizers. This is physically consistent with the blockage effects introduced by the centralizers. Finally, for the Class C slurry, the flow rate is 645.8 gpm. Velocity results are shown in Figures 5-14c ("15" means 1.5 in/sec) and 5-14d ("12" means 120 in/sec), and flowfield averages are given in Tables 5-12 and 5-13.

Taking n and k values typical of commonly used cement slurries, we have shown how velocity profiles for very general annular geometries can be computed in a stable and efficient manner. Also, the correct qualitative trends are captured; for example, increasing flow rate with eccentricity, decreasing volume flow with centralizer blockage. Simulation again allows us to answer "what if" questions quickly and inexpensively. What is the effect of a washout on bottom velocity? What effects will centralizer width have on velocity peaks? How does rheology interact with annular geometry at specific flow rates or pressure drops?

Again the primary operational concern in cementing is effective mud displacement and removal. Cement is known to channel through drilling mud, leaving mud behind. This may require expensive remedial work. But channeling is a hydrodynamic stability phenomenon that has been amply studied in the engineering literature. It is possible, as is routinely done in reservoir engineering as well as in outside industries, to evaluate the ability of any particular velocity profile to remain stable to flow disturbances. This robustness, which can be tailored by changing rheological properties or geometry, should be properly exploited to control undesirable fingering or channeling.

Figure 5-13a. Annular velocity, Class H slurry.

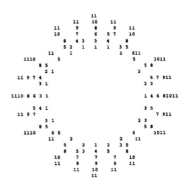

Figure 5-14b. Centralizer fitted mesh system.

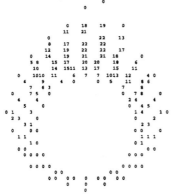

Figure 5-13b. Annular velocity, Class C slurry.

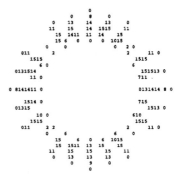

Figure 5-14c. Annular velocity, Class H slurry.

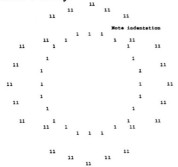

Figure 5-14a. Concentric casing and hole with centralizers.

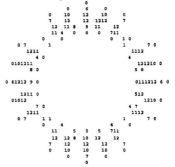

Figure 5-14d. Annular velocity, Class C slurry.

Table 5-12
Example 4: Summary, Average Quantities (Class H Slurry)

```
TABULATION OF CALCULATED AVERAGE QUANTITIES:
Area weighted means of absolute values taken over
BOTTOM HALF of annular cross-section ...
O  Average annular velocity = .9092E+00 in/sec
O  Average apparent viscosity = .1691E-02 lbf sec/sq in
O  Average stress, AppVis x dU/dx, = .1063E-02 psi
O  Average stress, AppVis x dU/dy, = .9636E-03 psi
O  Average dissipation = .4090E-02 lbf/(sec sq in)
O  Average shear rate dU/dx = .1682E+01 1/sec
O  Average shear rate dU/dy = .1447E+01 1/sec
O  Average Stokes product = .1840E-02 lbf/in

TABULATION OF CALCULATED AVERAGE QUANTITIES, II:
Area weighted means of absolute values taken over
ENTIRE annular (x,y) cross-section ...
O  Average annular velocity = .9185E+00 in/sec
O  Average apparent viscosity = .1674E-02 lbf sec/sq in
O  Average stress, AppVis x dU/dx, = .9290E-03 psi
O  Average stress, AppVis x dU/dy, = .9398E-03 psi
O  Average dissipation = .3970E-02 lbf/(sec sq in)
O  Average shear rate dU/dx = .1519E+01 1/sec
O  Average shear rate dU/dy = .1514E+01 1/sec
O  Average Stokes product = .1852E-02 lbf/in
```

Table 5-13
Example 4: Summary, Average Quantities (Class C Slurry)

```
TABULATION OF CALCULATED AVERAGE QUANTITIES:
Area weighted means of absolute values taken over
BOTTOM HALF of annular cross-section ...
O  Average annular velocity = .6993E+02 in/sec
O  Average apparent viscosity = .1812E-04 lbf sec/sq in
O  Average stress, AppVis x dU/dx, = .1206E-02 psi
O  Average stress, AppVis x dU/dy, = .1057E-02 psi
O  Average dissipation = .4154E+00 lbf/(sec sq in)
O  Average shear rate dU/dx = .1306E+03 1/sec
O  Average shear rate dU/dy = .1110E+03 1/sec
O  Average Stokes product = .1486E-02 lbf/in

TABULATION OF CALCULATED AVERAGE QUANTITIES, II:
Area weighted means of absolute values taken over
ENTIRE annular (x,y) cross-section ...
O  Average annular velocity = .7129E+02 in/sec
O  Average apparent viscosity = .1781E-04 lbf sec/sq in
O  Average stress, AppVis x dU/dx, = .1076E-02 psi
O  Average stress, AppVis x dU/dy, = .1078E-02 psi
O  Average dissipation = .4097E+00 lbf/(sec sq in)
O  Average shear rate dU/dx = .1181E+03 1/sec
O  Average shear rate dU/dy = .1177E+03 1/sec
O  Average Stokes product = .1494E-02 lbf/in
```

Example 5. Concentric Rotating Flows, Stationary Baseline

While cement slurry displaces drilling fluid in the annulus, the casing is often rotated from 10-20 rpm. The reasons are twofold: first to break the gel strength of any coagulated mud, and second to prevent flowing mud from gelling. The latter is amenable to flow simulation and modeling. It is of interest to compare the stress states for stationary versus rotating casings. We will assume a concentric geometry and use the analytical results of Chapter 3 obtained for rotating pipes. Again, the closed form expressions require a narrow annulus. The input summary shown in Table 5-14, extracted from output files, establishes reference conditions for a nonrotating flow run with "0.001 rpm." For stationary casings, the exact solution is of course available; however, for comparative reasons, we will not make use of it.

Table 5-14
Summary of Input Parameters

```
O Drillpipe outer radius (inches) =        4.0000
O Borehole radius (inches) =               5.0000
O Axial pressure gradient (psi/ft) =        .01000
O Drillstring rotation rate (rpm) =         .0010
O Drillstring rotation rate (rad/sec) =     .0001
O Fluid exponent "n" (nondimensional) =     .7240
O Consistency factor (lbf sec^n/sq in) =   .1861E-04
O Mass density of fluid (lbf sec^2/ft^4) = 1.9000
  (e.g., about 1.9 for water)
O Number of radial "grid" positions =          17
```

The software model calculates all physical quantities of possible interest. These include the variables discussed in Chapter 3, and, in addition, several derivative flow properties. The complete roster of output variables is given in Table 5-15; this same chart is supplied with all computed results.

Calculations for the foregoing quantities at each of the 17 control points selected in Table 5-14 require less than one second on Pentium machines. Various types of output are provided for correlation or research purposes, e.g., "high level" results such as,

```
Total volume flow rate (cubic in/sec) =  .2653E+03
                          (gal/min) =  .6889E+02
```

Or "low level results" conveniently supplied in hybrid text and ASCII plot form for complete portability; for example, Figure 5-15a for annular velocity or Figure 5-15b for the "maximum stress" defined in Table 5-15 next.

Table 5-15
Analytical (Non-Iterative) Solutions
Tabulated versus "r," Nomenclature and Units

r	Annular radial position ....….........	(in)
V_z	Velocity in axial z direction ....….	(in/sec)
V_θ	Circumferential velocity ...…..............	(in/sec)
$d\theta/dt$ or W	θ velocity ....….................	(rad/sec)
	(Note: 1 rad/sec = 9.5493 rpm)	
dV_z/dr	Velocity gradient ....…................	(1/sec)
dV_θ/dr	Velocity gradient ....…................	(1/sec)
dW/dr	Angular speed gradient ....…...........	(1/(sec $\times$ in))
$S_{r\theta}$	$r\theta$ stress component ....…................	(psi)
S_{rz}	rz stress component ......….................	(psi)
S_{max}	Sqrt ($S_{rz}**2 + S_{r\theta}**2$) ....….............	(psi)
dP/dr	Radial pressure gradient ...…...............	(psi/in)
App-Vis	Apparent viscosity ......…...................	(lbf sec /sq in)
Dissip	Dissipation function ......…...............	(lbf/(sec $\times$ sq in))
	(indicates frictional heat produced)	
Atan V_θ/V_z	Angle between V_θ and V_z vectors .....….	(deg)
Net Spd	Sqrt ($V_z**2 + V_\theta**2$) ......….................	(in/sec)
$D_{r\theta}$	$r\theta$ deformation tensor component ...…...	(1/sec)
D_{rz}	rz deformation tensor component ...…...	(1/sec)

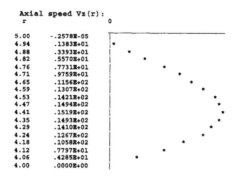

```
Axial speed Vz(r):
   r                       0
 5.00      -.2578E-05    ┌─────────────────────────
 4.94       .1383E+01    │*
 4.88       .3393E+01    │    *
 4.82       .5570E+01    │       *
 4.76       .7731E+01    │           *
 4.71       .9759E+01    │             *
 4.65       .1156E+02    │                *
 4.59       .1307E+02    │                  *
 4.53       .1421E+02    │                    *
 4.47       .1494E+02    │                     *
 4.41       .1519E+02    │                      *
 4.35       .1493E+02    │                     *
 4.29       .1410E+02    │                   *
 4.24       .1267E+02    │                 *
 4.18       .1058E+02    │              *
 4.12       .7797E+01    │         *
 4.06       .4285E+01    │     *
 4.00       .0000E+00    │
```

Figure 5-15a. Annular velocity.

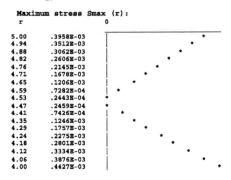

Maximum stress Smax (r):

r	0
5.00	.3958E-03
4.94	.3512E-03
4.88	.3062E-03
4.82	.2606E-03
4.76	.2145E-03
4.71	.1678E-03
4.65	.1206E-03
4.59	.7282E-04
4.53	.2443E-04
4.47	.2459E-04
4.41	.7426E-04
4.35	.1246E-03
4.29	.1757E-03
4.24	.2275E-03
4.18	.2801E-03
4.12	.3334E-03
4.06	.3876E-03
4.00	.4427E-03

Figure 5-15b. Maximum stress.

Detailed listings are also available for permanent reference; flow properties, tabulated against "r," can be read by spreadsheet programs for further trend analysis. A complete summary of computed results appears in Tables 5-16 to 5-19.

Table 5-16
Averaged Values of Annular Quantities

```
Average Vz (in/sec)  =  .9480E+01
           (ft/min)  =  .4740E+02
Average Vθ (in/sec)  =  .3253E-02
Average W (rad/sec)  =  .7401E-03
Average total speed (in/sec)  =  .9480E+01
Average angle between Vz and VÕ (deg)  =  .2633E+01
Average d(Vz)/dr (1/sec)  =  .0000E+00
Average d(Vθ)/dr (1/sec)  =  -.8717E-02

Average dW/dr (1/(sec × in))  =  -.2136E-02
Average dP/dr (psi/in)  =  .2709E-09
Average Srθ (psi)  =  .7355E-07
Average Srz (psi)  =  .7828E-05
Average Smax (psi)  =  .2097E-03
Average dissipation function (lbf sec^{n-2}/sq in) =.9225E-02
Average apparent viscosity (lbf sec^n/sq in)  =  .8692E-05
Average Drθ (1/sec)  =  .4729E-02
Average Drz (1/sec)  =  .8465E+00
```

Table 5-17. Detailed Tabulated Quantities.

r	Vz	Vθ	W	d(Vz)/dr	d(Vθ)/dr	dW/dr
5.00	-.258E-05	-.967E-08	-.193E-08	-.682E+02	-.101E-01	-.203E-02
4.94	.138E+01	.587E-03	.119E-03	-.578E+02	-.980E-02	-.201E-02
4.88	.339E+01	.115E-02	.236E-03	-.478E+02	-.941E-02	-.198E-02
4.82	.557E+01	.170E-02	.352E-03	-.383E+02	-.894E-02	-.193E-02
4.76	.773E+01	.221E-02	.465E-03	-.293E+02	-.838E-02	-.186E-02
4.71	.976E+01	.270E-02	.574E-03	-.209E+02	-.768E-02	-.175E-02
4.65	.116E+02	.316E-02	.680E-03	-.132E+02	-.678E-02	-.161E-02
4.59	.131E+02	.358E-02	.780E-03	-.658E+01	-.554E-02	-.138E-02
4.53	.142E+02	.395E-02	.873E-03	-.146E+01	-.340E-02	-.944E-03
4.47	.149E+02	.428E-02	.958E-03	.147E+01	-.344E-02	-.984E-03
4.41	.152E+02	.455E-02	.103E-02	.676E+01	-.585E-02	-.156E-02
4.35	.149E+02	.475E-02	.109E-02	.138E+02	-.752E-02	-.198E-02
4.29	.141E+02	.487E-02	.113E-02	.222E+02	-.895E-02	-.235E-02
4.24	.127E+02	.487E-02	.115E-02	.317E+02	-.103E-01	-.270E-02
4.18	.106E+02	.473E-02	.113E-02	.423E+02	-.116E-01	-.305E-02
4.12	.780E+01	.437E-02	.106E-02	.538E+02	-.129E-01	-.340E-02
4.06	.428E+01	.360E-02	.888E-03	.663E+02	-.144E-01	-.376E-02
4.00	.000E+00	.419E-03	.105E-03	.796E+02	-.164E-01	-.413E-02

Table 5-18. Detailed Tabulated Quantities.

r	Srθ	Srz	Smax	dP/dr	App-Vis	Dissip
5.00	.588E-07	-.396E-03	.396E-03	.171E-20	.580E-05	.270E-01
4.94	.602E-07	-.351E-03	.351E-03	.639E-11	.607E-05	.203E-01
4.88	.617E-07	-.306E-03	.306E-03	.250E-10	.640E-05	.146E-01
4.82	.632E-07	-.261E-03	.261E-03	.547E-10	.680E-05	.998E-02
4.76	.648E-07	-.214E-03	.214E-03	.943E-10	.733E-05	.628E-02
4.71	.664E-07	-.168E-03	.168E-03	.142E-09	.805E-05	.350E-02
4.65	.681E-07	-.121E-03	.121E-03	.197E-09	.913E-05	.159E-02
4.59	.699E-07	-.728E-04	.728E-04	.256E-09	.111E-04	.479E-03
4.53	.717E-07	-.244E-04	.244E-04	.316E-09	.168E-04	.356E-04
4.47	.736E-07	.246E-04	.246E-04	.376E-09	.167E-04	.361E-04
4.41	.756E-07	.743E-04	.743E-04	.430E-09	.110E-04	.502E-03
4.35	.776E-07	.125E-03	.125E-03	.475E-09	.901E-05	.172E-02
4.29	.798E-07	.176E-03	.176E-03	.506E-09	.791E-05	.390E-02
4.24	.820E-07	.227E-03	.227E-03	.514E-09	.717E-05	.722E-02
4.18	.843E-07	.280E-03	.280E-03	.492E-09	.662E-05	.118E-01
4.12	.868E-07	.333E-03	.333E-03	.426E-09	.619E-05	.179E-01
4.06	.893E-07	.388E-03	.388E-03	.293E-09	.585E-05	.257E-01
4.00	.919E-07	.443E-03	.443E-03	.402E-11	.556E-05	.353E-01

Table 5-19. Detailed Tabulated Quantities.

r	Vz	Vθ	Atan Vθ/Vz	Net Spd	Drθ	Drz
5.00	-.258E-05	-.967E-08	.215E+00	.258E-05	.507E-02	-.341E+02
4.94	.138E+01	.587E-03	.243E-01	.138E+01	.496E-02	-.289E+02
4.88	.339E+01	.115E-02	.195E-01	.339E+01	.482E-02	-.239E+02
4.82	.557E+01	.170E-02	.175E-01	.557E+01	.465E-02	-.191E+02
4.76	.773E+01	.221E-02	.164E-01	.773E+01	.442E-02	-.146E+02
4.71	.976E+01	.270E-02	.159E-01	.976E+01	.413E-02	-.104E+02
4.65	.116E+02	.316E-02	.157E-01	.116E+02	.373E-02	-.661E+01
4.59	.131E+02	.358E-02	.157E-01	.131E+02	.316E-02	-.329E+01
4.53	.142E+02	.395E-02	.159E-01	.142E+02	.214E-02	-.728E+00
4.47	.149E+02	.428E-02	.164E-01	.149E+02	.220E-02	.735E+00
4.41	.152E+02	.455E-02	.172E-01	.152E+02	.344E-02	.338E+01
4.35	.149E+02	.475E-02	.182E-01	.149E+02	.431E-02	.691E+01
4.29	.141E+02	.487E-02	.198E-01	.141E+02	.504E-02	.111E+02
4.24	.127E+02	.487E-02	.221E-01	.127E+02	.572E-02	.159E+02
4.18	.106E+02	.473E-02	.256E-01	.106E+02	.637E-02	.212E+02
4.12	.780E+01	.437E-02	.321E-01	.780E+01	.700E-02	.269E+02
4.06	.428E+01	.360E-02	.482E-01	.428E+01	.763E-02	.331E+02
4.00	.000E+00	.419E-03	.886E+02	.419E-03	.827E-02	.398E+02

Example 6. Concentric Rotating Flows, Rotating Casing

How do the computed results for non-rotating casing change if the casing were rotated at 20 rpm? Similar calculations were undertaken, with the following self-explanatory results in Tables 5-20 to 5-24.

Table 5-20
Summary of Input Parameters

```
O Drill pipe outer radius (inches)  =        4.0000
O Borehole radius (inches) =                 5.0000
O Axial pressure gradient (psi/ft) =          .01000
O Drillstring rotation rate (rpm) =         20.0000
O Drillstring rotation rate (rad/sec) =      2.0944
O Fluid exponent "n" (nondimensional) =       .7240
```
O Consistency factor (lbf $\sec^{n}$/sq in) = .1861E-04

O Mass density of fluid (lbf $\sec^2/ft^4$) =   1.9000
  (e.g., about 1.9 for water)
```
O Number of radial "grid" positions =           17
```

Table 5-21
Averaged Values of Annular Quantities

```
Average Vz (in/sec)  =   .1008E+02
           (ft/min)  =   .5040E+02
Average Vθ (in/sec)  =   .4987E+01
Average W (rad/sec)  =   .1146E+01
Average total speed (in/sec) =  .1157E+02

Average angle between Vz and Vθ (deg)  =   .2711E+02
Average d(Vz)/dr (1/sec)  =   .0000E+00

Average d(Vθ)/dr (1/sec) = -.1217E+02

Average dW/dr (1/(sec X in))  = -.3005E+01
Average dP/dr (psi/in)  =  .6528E-03

Average Srθ (psi)  =   .9562E-04
Average Srz (psi)  =   .7828E-05
Average Smax (psi)  =   .2386E-03
Average dissipation function (lbf sec^{n-2}/sq in)  =  .1073E-01
Average apparent viscosity (lbf sec^n/sq in)  =  .7458E-05
Average Drθ (1/sec)  =   .6656E+01
Average Drz (1/sec)  =   .9358E+00
```

Table 5-22. Detailed Tabulated Quantities.

r	Vz	Vθ	W	d(Vz)/dr	d(Vθ)/dr	dW/dr
5.00	-.397E-04	-.127E-04	-.253E-05	-.687E+02	-.133E+02	-.266E+01
4.94	.235E+01	.769E+00	.156E+00	-.584E+02	-.129E+02	-.263E+01
4.88	.459E+01	.151E+01	.310E+00	-.485E+02	-.124E+02	-.260E+01
4.82	.675E+01	.223E+01	.462E+00	-.390E+02	-.118E+02	-.255E+01
4.76	.883E+01	.292E+01	.612E+00	-.301E+02	-.112E+02	-.248E+01
4.71	.107E+02	.357E+01	.758E+00	-.218E+02	-.105E+02	-.238E+01
4.65	.124E+02	.418E+01	.900E+00	-.143E+02	-.963E+01	-.227E+01
4.59	.138E+02	.476E+01	.104E+01	-.787E+01	-.878E+01	-.214E+01
4.53	.149E+02	.528E+01	.117E+01	-.246E+01	-.821E+01	-.207E+01
4.47	.155E+02	.576E+01	.129E+01	.250E+01	-.843E+01	-.217E+01
4.41	.157E+02	.619E+01	.140E+01	.820E+01	-.945E+01	-.246E+01
4.35	.153E+02	.656E+01	.151E+01	.152E+02	-.108E+02	-.283E+01
4.29	.144E+02	.687E+01	.160E+01	.235E+02	-.123E+02	-.323E+01
4.24	.129E+02	.713E+01	.168E+01	.330E+02	-.138E+02	-.365E+01
4.18	.108E+02	.736E+01	.176E+01	.435E+02	-.153E+02	-.407E+01
4.12	.793E+01	.760E+01	.184E+01	.550E+02	-.167E+02	-.451E+01
4.06	.435E+01	.791E+01	.195E+01	.674E+02	-.182E+02	-.497E+01
4.00	.000E+00	.838E+01	.209E+01	.807E+02	-.197E+02	-.545E+01

Table 5-23. Detailed Tabulated Quantities.

r	Srθ	Srz	Smax	dP/dr	App-Vis	Dissip
5.00	.765E-04	-.396E-03	.403E-03	.294E-14	.576E-05	.282E-01
4.94	.783E-04	-.351E-03	.360E-03	.110E-04	.602E-05	.215E-01
4.88	.802E-04	-.306E-03	.316E-03	.430E-04	.632E-05	.159E-01
4.82	.822E-04	-.261E-03	.273E-03	.944E-04	.668E-05	.112E-01
4.76	.842E-04	-.214E-03	.230E-03	.163E-03	.713E-05	.744E-02
4.71	.863E-04	-.168E-03	.189E-03	.248E-03	.769E-05	.463E-02
4.65	.885E-04	-.121E-03	.150E-03	.345E-03	.841E-05	.266E-02
4.59	.908E-04	-.728E-04	.116E-03	.452E-03	.925E-05	.147E-02
4.53	.932E-04	-.244E-04	.964E-04	.565E-03	.994E-05	.934E-03
4.47	.957E-04	.246E-04	.988E-04	.681E-03	.985E-05	.991E-03
4.41	.982E-04	.743E-04	.123E-03	.795E-03	.905E-05	.168E-02
4.35	.101E-03	.125E-03	.160E-03	.905E-03	.819E-05	.314E-02
4.29	.104E-03	.176E-03	.204E-03	.101E-02	.747E-05	.557E-02
4.24	.107E-03	.227E-03	.251E-03	.110E-02	.690E-05	.915E-02
4.18	.110E-03	.280E-03	.301E-03	.119E-02	.644E-05	.140E-01
4.12	.113E-03	.333E-03	.352E-03	.128E-02	.607E-05	.204E-01
4.06	.116E-03	.388E-03	.405E-03	.141E-02	.575E-05	.285E-01
4.00	.120E-03	.443E-03	.459E-03	.161E-02	.549E-05	.383E-01

Table 5-24. Detailed Tabulated Quantities.

r	Vz	Vθ	Atan Vθ/Vz	Net Spd	Drθ	Drz
5.00	-.397E-04	-.127E-04	.177E+02	.416E-04	.664E+01	-.344E+02
4.94	.235E+01	.769E+00	.181E+02	.247E+01	.651E+01	-.292E+02
4.88	.459E+01	.151E+01	.183E+02	.483E+01	.635E+01	-.242E+02
4.82	.675E+01	.223E+01	.183E+02	.711E+01	.615E+01	-.195E+02
4.76	.883E+01	.292E+01	.183E+02	.930E+01	.591E+01	-.150E+02
4.71	.107E+02	.357E+01	.184E+02	.113E+02	.561E+01	-.109E+02
4.65	.124E+02	.418E+01	.186E+02	.131E+02	.527E+01	-.717E+01
4.59	.138E+02	.476E+01	.190E+02	.146E+02	.491E+01	-.394E+01
4.53	.149E+02	.528E+01	.195E+02	.158E+02	.469E+01	-.123E+01
4.47	.155E+02	.576E+01	.204E+02	.166E+02	.486E+01	.125E+01
4.41	.157E+02	.619E+01	.215E+02	.169E+02	.543E+01	.410E+01
4.35	.153E+02	.656E+01	.231E+02	.167E+02	.616E+01	.761E+01
4.29	.144E+02	.687E+01	.254E+02	.160E+02	.694E+01	.118E+02
4.24	.129E+02	.713E+01	.289E+02	.148E+02	.772E+01	.165E+02
4.18	.108E+02	.736E+01	.343E+02	.130E+02	.851E+01	.217E+02
4.12	.793E+01	.760E+01	.438E+02	.110E+02	.929E+01	.275E+02
4.06	.435E+01	.791E+01	.612E+02	.902E+01	.101E+02	.337E+02
4.00	.000E+00	.838E+01	.900E+02	.838E+01	.109E+02	.404E+02

Observe that our calculated "maximum stresses" S_{max} have increased significantly from those of Example 5 without rotation. This increase may thwart flowing mud from gelling, that is, prevent the cement slurry from channeling and bypassing isolated pockets of resistive gelled mud. The closed form solutions provide a means to estimate quickly changes to flowing properties, and are valuable in this respect. Again, the model predicts total volume flow rate. In the present case, the result

```
Total volume flow rate (cubic in/sec) = .2828E+03
                         (gal/min) = .7345E+02
```

exceeds that obtained for non-rotating casing of Example 5 under the same pressure drop. This is physically consistent with well-known effects of rotation. Detailed plots are available for any of the tabulated quantities, for example, Figures 5-16a and 5-16b for velocity and stress.

```
Axial speed Vz(r):
  r                      0
5.00    -.3966E-04        |
4.94     .2350E+01        |  *
4.88     .4586E+01        |    *
4.82     .6754E+01        |      *
4.76     .8827E+01        |        *
4.71     .1074E+02        |          *
4.65     .1243E+02        |            *
4.59     .1383E+02        |              *
4.53     .1488E+02        |                *
4.47     .1551E+02        |                 *
4.41     .1568E+02        |                  *
4.35     .1534E+02        |                 *
4.29     .1444E+02        |               *
4.24     .1293E+02        |             *
4.18     .1077E+02        |           *
4.12     .7926E+01        |        *
4.06     .4348E+01        |     *
4.00     .0000E+00        |
```

Figure 5-16a. Annular velocity.

```
Maximum stress Smax (r):
  r                      0
5.00     .4032E-03        |                        *
4.94     .3599E-03        |                      *
4.88     .3165E-03        |                    *
4.82     .2732E-03        |                  *
4.76     .2304E-03        |                *
4.71     .1887E-03        |              *
4.65     .1496E-03        |            *
4.59     .1164E-03        |          *
4.53     .9636E-04        |         *
4.47     .9879E-04        |         *
4.41     .1232E-03        |          *
4.35     .1604E-03        |            *
4.29     .2040E-03        |              *
4.24     .2512E-03        |                *
4.18     .3007E-03        |                  *
4.12     .3520E-03        |                    *
4.06     .4046E-03        |                      *
4.00     .4586E-03        |                         *
```

Figure 5-16b. Maximum stress.

Coiled Tubing Return Flows

After a well has been drilled and cemented, and after it has seen production, fines and sands may emerge through the perforations. This may be the case with unconsolidated sands and unstable wellbores; in deviated and horizontal wells, the debris remains on the low side of the hole. To remove these fines, metal tubing unrolled from "coils" at the surface (typically, with 1 to 2 inches, outer diameter) is run to the required depth. Fluids are pumped downhole through this tubing. They clean the highly eccentric annulus, and return to the surface carrying the debris.

The clean-up process is not unlike cuttings transport in drilling, except that hole eccentricities here are more severe. Besides sand washing, coiled tubing is also used in paraffin cleanout, acid or cement squeezes, and mud displacement. Typical fluids may include nitrogen or non-Newtonian foams. Figure 5-17a displays a typical annulus encountered in coiled tubing applications; note the large diameter ratio and the typically high eccentricities.

The eccentric flow model of Chapter 2 can be used to simulate fluid motions in such annuli. Figures 5-17b and 5-17c, for example, display the boundary conforming mesh generated for this system and a typical velocity field computed in these coordinates. As in cuttings transport, velocity and stress are expected to play important roles in sand cleaning. Annular flow simulation, again, is straightforward and robust; the eccentric model calculates the required quantities accurately. Note how no-slip velocity boundary conditions are exactly satisfied at all solid surfaces.

Heavily Clogged Stuck Pipe

So far we have considered *annular* flows only, that is, dough-nut-shaped "doubly-connected" geometries. How can we model "singly-connected" *pipe-like* geometries, not atypical of annuli highly clogged by thick cuttings beds, without completely reformulating the problem? Figure 5-18a shows one possible clogged configuration, where we have displayed the *drillpipe plus cuttings bed* as a single entity whose boundary is marked by asterisks.

At the lower annulus, the separation from the borehole bottom is kept to a token minimum, say 0.01 inch. The computer program will give nearly zero velocities for this narrow gap since no-slip conditions predominate. For all practical purposes, the gap is impermeable to flow and completely plugs up the bottom. Hence, the only flow domain of dynamical significance is the simple region just above the thick cuttings bed. A typical mesh system and a velocity field are shown in Figures 5-18b and 5-18c for a power law mud.

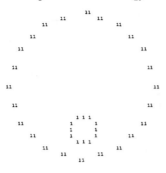

Figure 5-17a. Highly eccentric annulus.

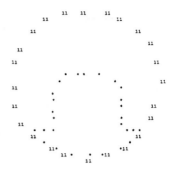

Figure 5-18a. Heavily clogged annulus.

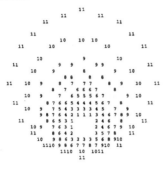

Figure 5-17b. Coiled tubing mesh.

Figure 5-18b. Computed mesh.

Figure 5-17c. Annular velocity.

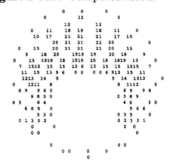

Figure 5-18c. Annular velocity.

CONCLUSIONS

We have approached the subject of annular flow quite generally in Chapters 2, 3, and 4, and developed models for different geometries in several physical limits for Newtonian and non-Newtonian rheologies. These models were used in the present chapter to study several drilling and production problems of importance in deviated and horizontal wells. These included cuttings transport, spotting fluid analysis, cementing, and coiled tubing applications. In each case, our limited studies addressed only the annular flow simulation aspects of the problem. We claim only that the models developed here are useful in correlating real world behavior with flow properties that, until now, have resisted analysis, and that our studies appear to have identified some potentially useful ideas.

With respect to our eccentric flow solutions, we showed here and in Chapter 2 that the numerical model can be run "as is" and still *generate physically correct qualitative trends and quantitative estimates.* Mesh dependence, of importance at least academically, is not significant here. This robustness is very important from a practical and operational viewpoint since few exact solutions are ever available for mesh calibration.

We emphasize that the use of the rotational viscometer, or any other mechanical viscometer, should be restricted to determining the *intrinsic* rheological parameters (such as "n" or "k") that characterize a particular non-Newtonian fluid. Well-designed viscometers are ideal for this purpose. They are, however, not suitable for predicting actual downhole properties during specific drilling and cementing runs because actual shear rates, which can and do vary widely, are always unknown a priori: they depend on the details of the annular geometry. This explains why conventional attempts to correlate cuttings transport data with viscometer properties have been unsuccessful.

A viscometer that actually predicts downhole properties is, of course, difficult to construct. Given all the nonlinearities and nuances of non-Newtonian flow and borehole eccentricity, it is unduly optimistic to believe that a dimensionally scalable mechanical device can be built at all. But that is not necessary. As we have amply demonstrated, software simulation provides an elegant and efficient means to achieve most drilling and production objectives. And it can be improved upon as new techniques and insights materialize.

We have dealt only with the drilling and production aspects of annular flow. These of course represent small subsets of industrial application. Annular flow is important in chemical engineering, manufacturing and extrusion processes. It is clear that the methods developed in this book can be applied to disciplines outside the petroleum industry, and an effort is underway to explore these possibilities.

REFERENCES

Adewumi, M.A., and Tian, S., "Hydrodynamic Modeling of Wellbore Hydraulics in Air Drilling," SPE Paper No. 19333, 1989 SPE Eastern Regional Meeting, Morgantown, October 24-27, 1989.

Baret, F., Free, D.L., and Griffin, T.J., "Hard and Fast Rules for Effective Cementing," *Drilling Magazine*, March-April 1989, pp. 22-25.

Becker, T.E., Azar, J.J., and Okrajni, S.S., "Correlations of Mud Rheological Properties with Cuttings Transport Performance in Directional Drilling," SPE Paper No. 19535, 64th Annual Technical Conference and Exhibition, Society of Petroleum Engineers, San Antonio, October 1989.

Benge, G., "Field Study of Offshore Cement-Spacer Mixing," *SPE Drilling Engineering*, September 1990, pp. 196-200.

Brown, N.P., Bern, P.A., and Weaver, A., "Cleaning Deviated Holes: New Experimental and Theoretical Studies," SPE/IADC Paper No. 18636, 1989 SPE/IADC Drilling Conference, New Orleans, February 28 - March 3, 1989.

Chin, W.C., "Advances in Annular Borehole Flow Modeling," *Offshore Magazine*, February 1990, pp. 31-37.

Chin, W.C., "Exact Cuttings Transport Correlations Developed for High Angle Wells," *Offshore Magazine*, May 1990, pp. 67-70.

Chin, W.C., "Annular Flow Model Explains Conoco's Borehole Cleaning Success," *Offshore Magazine*, October 1990, pp. 41-42.

Chin, W.C., "Model Offers Insight into Spotting Fluid Performance," *Offshore Magazine*, February 1991, pp. 32-33.

Chin, W.C., "Eccentric Annular Flow Modeling for Highly Deviated Boreholes," *Offshore Magazine*, August 1993.

Ford, J.T., Peden, J.M., Oyeneyin, M.B., Gao, E., and Zarrough, R., "Experimental Investigation of Drilled Cuttings Transport in Inclined Boreholes," SPE Paper No. 20421, 65th Annual Technical Conference and Exhibition of the Society of Petroleum Engineers, New Orleans, September 23-26, 1990.

Fraser, L.J., "Field Application of the All-Oil Drilling Fluid Concept," IADC/SPE Paper No. 19955, 1990 IADC/SPE Drilling Conference, Houston, February 27-March 2, 1990.

Fraser, L.J., "Green Canyon Drilling Benefits from All Oil Mud," *Oil and Gas Journal*, March 19, 1990, pp. 33-39.

Fraser, L.J., "Effective Ways to Clean and Stabilize High-Angle Hole," *Petroleum Engineer International*, November 1990, pp. 30-35.

George, C., "Innovations Change Cementing Operations," *Petroleum Engineering International*, October 1990, pp. 37- 41.

Gray, K.E., "The Cutting Carrying Capacity of Air at Pressures Above Atmosphere," *Petroleum Transactions*, AIME, Vol. 213, 1958, pp. 180-185.

Halliday, W.S., and Clapper, D.K., "Toxicity and Performance Testing of Non-Oil Spotting Fluid for Differentially Stuck Pipe," Paper No. 18684, SPE /IADC Drilling Conference, New Orleans, February 28-March 3, 1989.

Harvey, F., "Fluid Program Built Around Hole Cleaning Protecting Formation," *Oil and Gas Journal*, 5 November 1990, pp. 37-41.

Hussaini, S.M., and Azar, J.J., "Experimental Study of Drilled Cuttings Transport Using Common Drilling Muds," *Society of Petroleum Engineers Journal*, February 1983, pp. 11-20.

Lin, C.C., *The Theory of Hydrodynamic Stability*, Cambridge University Press, London, 1967.

Lockyear, C.F., Ryan, D.F., and Gunningham, M.M., "Cement Channeling:How to Predict and Prevent," *SPE Drilling Engineering*, September 1990, pp. 201-208.

Martin, M., Georges, C., Bisson, P., and Konirsch, O., "Transport of Cuttings in Directional Wells," SPE/IADC Paper No. 16083, 1987 SPE/IADC Drilling Conference, New Orleans, March 15-18, 1987.

Okragni, S.S., "Mud Cuttings Transport in Directional Well Drilling," SPE Paper No. 14178, 60th Annual Technical Conference and Exhibition of the Society of Petroleum Engineers, Las Vegas, September 22-25, 1985.

Outmans, H.D., "Mechanics of Differential Pressure Sticking of Drill Collars," *Petroleum Transactions*, AIME, Vol. 213, 1958, pp. 265-274.

Quigley, M.S., Dzialowski, A.K., and Zamora, M., "A Full-Scale Wellbore Friction Simulator," IADC/SPE Paper No. 19958, 1990 IADC/SPE Drilling Conference, Houston, February 27-March 2, 1990.

Savins, J.G., and Wallick, G.C., "Viscosity Profiles, Discharge Rates, Pressures, and Torques for a Rheologically Complex Fluid in a Helical Flow," *A.I.Ch.E.Journal*, Vol. 12, No. 2, March 1966, pp. 357-363.

Seeberger, M.H., Matlock, R.W., and Hanson, P.M., "Oil Muds in Large Diameter, Highly Deviated Wells: Solving the Cuttings Removal Problem," SPE/IADC Paper No. 18635, 1989 SPE/IADC Drilling Conference, New Orleans, February 28-March 3, 1989.

Seheult, M., Grebe, L., Traweek, J.E., and Dudley, M., "Biopolymer Fluids Eliminate Horizontal Well Problems," *World Oil*, January 1990, pp. 49-53.

Sifferman, T.R., and Becker, T.E., "Hole Cleaning in Full-Scale Inclined Wellbores," SPE Paper No. 20422, 65th Annual Technical Conference and Exhibition of the Society of Petroleum Engineers, New Orleans, September 23-26, 1990.

Smith, D.K., *Cementing*, Society of Petroleum Engineers, Dallas, 1976.

Smith, T.R., "Cementing Displacement Practices - Field Applications," *Journal of Petroleum Technology*, May 1990, pp. 564-566.

Sparlin, D.D., and Hagen, R.W., "Controlling Sand in a Horizontal Completion," *World Oil*, November 1988, pp. 54-58.

Streeter, V.L., *Handbook of Fluid Dynamics*, McGraw-Hill, New York, 1961.

Suman, G.O., "Cementing - The Drilling/Completion Interface," Completion Tool Company Technical Report, Houston, September 1988.

Suman, G.O., *Cementing*, Completion Tool Company, Houston, 1990.

Suman, G.O., and Ellis, R.C., *World Oil's Cementing Handbook*, Gulf Publishing Company, Houston, 1977.

Suman, G.O., and Snyder, R.E., "Primary Cementing: Why Many Conventional Jobs Fail," *World Oil*, December 1982.

Tomren, P.H., Iyoho, A.W., and Azar, J.J., "Experimental Study of Cuttings Transport in Directional Wells," *SPE Drilling Engineering*, February 1986, pp. 43-56.

Wilson, M.A., and Sabins, F.L., "A Laboratory Investigation of Cementing Horizontal Wells," *SPE Drilling Engineering*, September 1988, pp. 275-280.

Zaleski, T.E., and Ashton, J.P., "Gravel Packing Feasible in Horizontal Well Completions," *Oil and Gas Journal*, 11 June 1990, pp. 33-37.

Zaleski, T.E., and Spatz, E., "Horizontal Completions Challenge for Industry," *Oil and Gas Journal*, 2 May 1988, pp. 58-70.

6

Bundled Pipelines:
Coupled Annular Velocity and Temperature

Our computational rheology efforts in the early 1990s focused on the physics of debris removal in annular flow, in particular, highly eccentric geometries containing non-Newtonian fluids. To support this objective, fast, stable, and robust algorithms were designed to map irregular domains to rectangular ones, where the nonlinear partial differential flow equations were solved accurately. Because analytical solutions were not available for validation, we turned to empirical data to establish the physical consistency of results, and in the process, developed computational tools that were useful in drilling and production. Our "ASCII text plots," for example, as shown in Figure 6-1a, overlaid axial velocity and derived quantities like apparent viscosity, shear rate, and viscous stress on the borehole geometries themselves, and were invaluable in facilitating the benchmarking process. They provided highly visual information that bore quantitative value, allowing engineers to understand quickly the mechanics of particular flows. Before pursuing the subject of this chapter, we will review recent advances made in *interpreting* computed flowfields that enable us to understand results much more accurately.

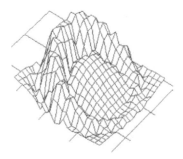

Figure 6-1a. Velocity distribution, numerical text plot.

Figure 6-1b. Velocity surface plot, simple "mesh" diagram.

159

COMPUTER VISUALIZATION AND SPEECH SYNTHESIS

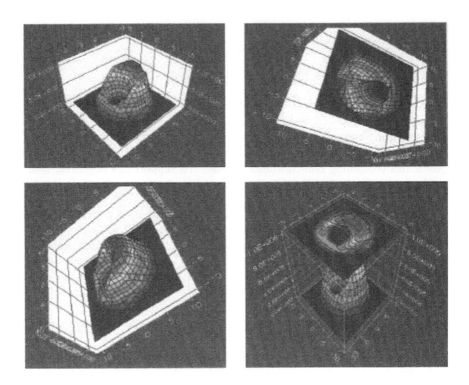

Figure 6-2. Three-dimensional, "dynamic" plots (see CDROM for color slides).

In the early 1990s, slow 8086, 286, and 386 Intel processors were standard; fast, high-resolution color plotting and graphical user interfaces were not commonly available, and the Windows operating system was not widely used. The software complementary to *Borehole Flow Modeling*, forerunner to this volume, was based on MS-DOS, drawing on portable text graphics only and "mesh plots" from independent commercial packages.

Computer visualization.   Hardware and software costs have, in the meantime, declined exponentially, and the graphical environment has changed significantly.   While numerically oriented velocity plots, e.g., as shown in Figure 6-1a, are still valuable in providing a quantitative "feel" for the physics, these are now augmented by "dynamic plots" such as those in Figure 6-2. These plots, in addition to assigning color scales to the velocity distribution, also associate a height attribute to it.   The resulting diagram can be "interrogated" using a computer mouse, for example, taking "translate, rotate, and zoom" actions. In addition, contour diagrams are readily superposed on the three-dimensional structure. "Flat plots" are also available, as shown in the sequence of diagrams in Figures 6-3a,b, for all of the properties introduced in Chapter 2. To enhance report generation capabilities, the software is now fully Windows compatible. Thus, files and graphics created by the new software can be easily "pasted" into standard word processing and slide presentation programs.

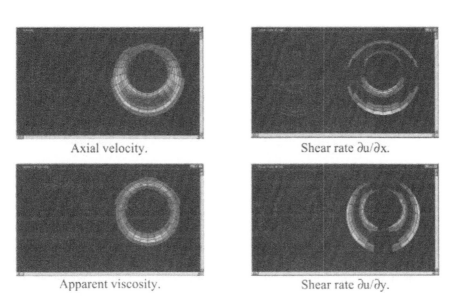

| Axial velocity. | Shear rate $\partial u/\partial x$. |
| Apparent viscosity. | Shear rate $\partial u/\partial y$. |

Figure 6-3a. Flat planar color plots (see CDROM for color slides).

Physical processes such as cuttings transport in boreholes, and particularly, wax deposition and hydrate formation in cold subsea pipelines, depend on properties like flow velocity, viscous stress, and field temperature. For example, the buildup and erosion of solids beds depend on velocity when the constituent particles are discrete and noncohesive; on the other hand, when they are cohesively formed, stress provides a more realistic correlation parameter.

In any event, workers in computational rheology encounter voluminous numerical data, making integrated color graphics a necessary part of any automated analysis capability. But the end objective in output processing is not, in itself, computer visualization, although this remains an integral part of the output. Very clearly, advanced processing must support *interpretation*, which in turn improves engineering productivity and increases system reliability.

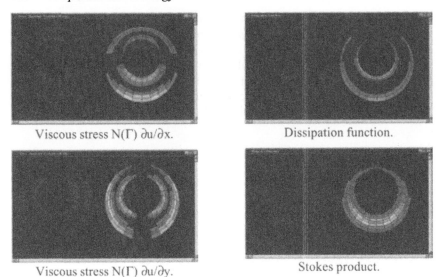

Viscous stress N(Γ) ∂u/∂x. Dissipation function.

Viscous stress N(Γ) ∂u/∂y. Stokes product.

Figure 6-3b. Flat planar color plots (see CDROM for color slides).

Speech synthesis. Three-dimensional, color visualization helps users identify trends in computed data, for example, highlighting increases and decreases in viscous stress within the borehole annulus. However, it does not assist the novice in *interpretation* as would a skilled teacher in a controlled classroom setting. Such an instructor might warn of impending cuttings transport problems when the flowing stream contains a high percentage of drilling debris, or perhaps, suggest remedial action that would prevent operational hazards. "Speech synthesis" was viewed as a potentially important capability in output processing. In the CDROM distributed with this book, "text-to-speech" features are provided that literally read text files aloud to the user; in continuing development prototypes, an "intelligent" intepretation algorithm that *explains* visual outputs to users is being refined. Readers interested in acquiring this technology, hosted on standard personal computers, should contact the author directly.

COUPLED VELOCITY AND TEMPERATURE FIELDS

So far in this book, we have assumed that annular velocity can be computed independently of temperature, and for the great majority of problems encountered in drilling and cementing, this is the case. In deep subsea applications, this may not be true. Hot produced crude may contain dissolved waxes, which precipitate out of solution when the pipeline has cooled sufficiently. In addition, the environment may be conducive to hydrate plug formation. Thus, there is a practical need to heat the produced fluid so that it can be efficiently transported to market. Numerous methods have been suggested to "bundle" pipelines together, but for the purposes of discussion, we will consider a particularly simple concept in order to illustrate the mathematical modeling issues and computationat techniques that apply.

Simple bundle concept. In this chapter, we will consider the simple bundled pipeline concept shown in Figure 6-4. As before, fluid flows axially "out of the page" within the "large circle" below, with the "small circle" defining a region of space held at a much higher temperature. In practice, this "heat pipe" will contain hot fluid also flowing axially, but this inner flow is completely isolated from the annular flow just outside. The dark rectangular region surrounding both circles represents the much colder ocean, which for all practical purposes, is infinite in extent.

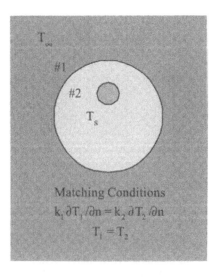

Figure 6-4. Detailed temperature model.

In mathematical modeling, progress is made by discarding as many terms in the governing equations as possible, while retaining the essential elements of the physics. For example, in Chapter 2, we assumed that fluid flowed unidirectionally along the axis of the borehole; thus, "u" was nonzero, while the transverse velocity components "v" and "w" vanished identically. This led to a single partial differential equation, which was nonetheless difficult to solve because it was nonlinear and because the flow domain was complicated.

In the present problem, the fluid-dynamical equations are extremely complicated unless several assumptions are made. For instance, in heat transfer, "free" or "natural convection" is often modeled using the simplified "Boussinesq approximation," that is, the effects of temperature differences are simulated only through spatially dependent buoyancy in the body force term and not compressibility. But even this simplification leads to difficulty. If a vertical flow "v" is permitted in Figure 6-4 in addition to our axial flow "u," it is clear that the remaining transverse velocity component "w" must be nonzero in order for the fluid to "turn" within the annular cross-section.

Thus, all three (coupled) momentum equations must be considered, together with the complete mass continuity equation; and since viscosity is temperature dependent, there is further coupling with the energy equation. These difficulties do not mean that modeling is not possible: it is, but the resulting algorithms will require high-power workstations or mainframes.

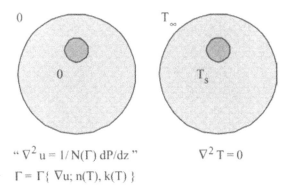

"$\nabla^2 u = 1/N(\Gamma) \, dP/dz$" $\nabla^2 T = 0$

$\Gamma = \Gamma\{ \nabla u; n(T), k(T) \}$

Figure 6-5a,b. Simplified velocity and thermal model.

For high production rates, the effects of free convection are less significant, and the assumption of purely axial flow may suffice. Thus, the momentum model developed in Chapter 2 for non-Newtonian flows can be used, provided we understand that rheological properties like "n(T)" and "k(T)" now depend on temperature T, which will vary about the flow annulus shown in Figure 6-5.

The more complicated cross-sectional thermal boundary condition model in Figure 6-4, which matches temperature and normal heat flux at the annulus and ocean interface, can be replaced by the simpler model shown at the right of Figure 6-5. If, in the complete energy equation, we neglect changes in the axial direction, transient effects, and internal heat generation, Laplace's equation for T(y,x) arises in the simple limit of constant thermal properties. This leads to the elementary pipe bundle model summarized in Figure 6-5.

In this computational scheme, temperature is solved first, and stored as a function of (y,x). Then, the flow solver developed in Chapter 2 is used, but "n" and "k" now vary with position. Importantly, both u(y,x) and T(y,x) are solved by the *same* "boundary conforming, curvilinear mesh" algorithms developed previously since the governing equations are almost identical. Of course, more complicated physical flows do exist, which require more elaborate models; other bundling concepts may involve multiple embedded pipes, resulting in annuli with multiple "holes." These extensions will be discussed later.

Design issues. If we assume that the pipeline is flowing under a fixed pressure gradient, that is, acting through reservoir pressure alone, it is clear that under isothermal conditions, the volume flow rate for the annulus in Figure 6-5a in general increases with increasing eccentricity, as demonstrated in Chapter 2. Of course, the temperature field T(y,x) in reality varies throughout, so that the rheological properties of the fluid likewise vary. In general, n, k, and yield stress, or other properties characterizing the constitutive relation, will depend on temperature and wax content, in a manner that must be determined in the laboratory. Once these relations are obtained, several design issues arise.

For example, the total revenue to the operator depends on the net volume of fluid delivered; in this respect, qualitatively speaking, eccentricity is good. However, high eccentricity may imply that the surrounding fluid is not uniformly heated, so that waxes and hydrates form locally and possibly precipitate out of solution. Centralization of the heating pipe may suggest less throughput, in the sense that there is less eccentricity; bear in mind, though, that increased temperatures imply reduced viscosity, which lead to more efficient production. Thus, tradeoff studies are required, based on detailed simulation. Cost minimization should be planned at the outset. Fixed costs include those associated with the sizes of inner and outer pipes, and the purchase of heating systems. Variable elements are related to energy costs, "down times" due to "plugging" or required maintenance, and so on. In general, it is clear that the economic optimization problem is very nonlinear, and must include the coupled velocity and thermal simulator as an essential element.

Sample calculation. To illustrate the basic approach, we might consider a flowline having a given diameter, in one case containing a narrow heat pipe, and in the second, a wider one. Figure 6-6a shows the computed velocity field for the narrow pipe, with high "red" velocities in the eccentric part of the annulus, while Figure 6-7a illustrates the analogous velocity result for the much wider one. Figures 6-6b and 6-7b display solutions for the calculated temperature fields, which are high at the inner pipe and cold at the edge of the annulus. Computed results are sensitive to input parameters, and general conclusions cannot be given. Total volume flow rates can be computed from the velocity solution, while total heat flux, needed for energy cost calculations, can be obtained from field results for calculated temperature.

Over the life of the well, different crudes with varying viscosities and volume flow rates can be produced. What if the operator wished to increase the source temperature T_s in Figure 6-5b in order to decrease the average viscosity? Is it necessary to consider the heat transfer boundary value problem "off-line"? It turns out that, while we have alluded to its solution using methods already devised, solutions to Laplace's equation are *automatically* provided by the mapping function! This unique property is discussed in a later chapter.

The modeling discussed above focuses on computational methods for eccentric annular domains. Other approaches dealing with different aspects of the physics may also be relevant to practical design, but are not discussed here, given the "boundary conforming mesh" orientation of this book. The researches cited in the references provide a cross-section of professional interests.

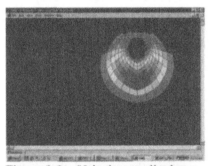

Figure 6-6a. Velocity, small tube.

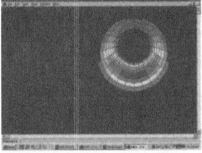

Figure 6-7a. Velocity, large tube.

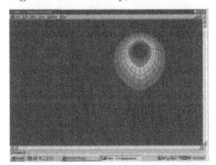

Figure 6-6b. Temperature, small tube.

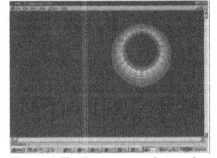

Figure 6-7b. Temperature, large tube.

Figures 6, 7. Comparing two bundled pipeline concepts (see CDROM).

REFERENCES

Brown, T.S., Clapham, J., Danielson, T.J., Harris, R.G., and Erickson, D.D., "Application of a Transient Heat Transfer Model for Bundled, Multiphase Pipelines," SPE Paper No. 36610, Society of Petroleum Engineers Annual Technical Conference and Exhibition, Denver, 1996.

Danielson, T.J., and Brown, L.D., "An Analytical Model for a Flowing Bundle System," SPE Paper No. 56719, Society of Petroleum Engineers Annual Technical Conference and Exhibition, Houston, 1999.

Zabaras, G.J., and Zhang, J.J., "Bundle Flowline Thermal Analysis," SPE Paper No. 52632, Society of Petroleum Engineers Annual Technical Conference and Exhibition, San Antonio, 1997.

7

Pipe Flow Modeling in General Ducts

In this chapter, we model steady, laminar pipe flow in straight ducts containing general non-Newtonian fluids, importantly, allowing arbitrary cross-sections, while satisfying exact "no-slip" velocity boundary conditions. We envision a duct taking the general form in Figure 7-1, for which we will calculate the axial velocity distribution everywhere; from this, we determine physical quantities such as apparent viscosity, shear rate, and viscous stress. In developing a general solver, first we will need to understand the strengths and limitations of existing solution methods, so that we can extend their solution attributes in directions that impose the fewest engineering restrictions. That is, we seek to model the most general rheologies in arbitrarily shaped ducts. We will review classical linear techniques first, and then apply them to idealized duct geometries such as circles and rectangles containing Newtonian flow. The solutions to these simpler problems also provide quantitative results required for later benchmarking and validation, key ingredients needed to ensure that our general approach does indeed predict physically correct results and trends.

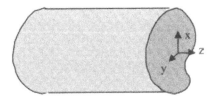

Figure 7-1. General duct.

Again, we begin with "simple" Newtonian flow, and focus on two important problems first. If "z" represents the axial direction and dp/dz is the constant pressure gradient along z, the axial velocity distribution "u" satisfies the equation $\nabla^2 u = 1/\mu \; dp/dz$, where μ is a constant viscosity. Note that the assumption of straight flow simplifies the original Navier-Stokes equations by eliminating the nonlinear convective terms, leaving a reduced equation having classical Poisson form. In this equation, "∇^2" is the Laplacian in the cross-flow plane, and takes on different forms for different coordinate systems.

For example, in the case of circular pipes, $\nabla^2 = d^2/dr^2 + 1/r\, d/dr$ where "r" is the cylindrical radial coordinate. For rectangular ducts, $\nabla^2 = \partial^2/\partial x^2 + \partial^2/\partial y^2$ applies, where "x" and "y" are the usual Cartesian variables. Here, we will develop in detail the ideas first presented in Chapter 2, and show that the boundary conforming "r-s" mesh systems introduced earlier are the natural coordinates with which to express "∇^2" for the geometry shown in Figure 7-1.

NEWTONIAN FLOW IN CIRCULAR PIPES

The classical solution for Hagen-Poiseuille flow through a pipe is a well-known formula used by fluid-dynamicists. In this section, we will derive it from first principles, and show how, despite the apparent generality, the solution methods are really quite restrictive. We will also show how complementary numerical solutions can be constructed, and take the opportunity to introduce the finite difference methods so widely used in the process and piping industry. These methods are, in fact, used to solve the grid generation equations, and also, the nonlinear flow equations written to the transformed mesh.

Exact analytical solution. The classical solution for steady, laminar, Newtonian flow in an infinite circular pipe, as shown in Figure 7-2, without roughness, turbulence, or inlet entry effects, is obtained by solving

$$d^2u(r)/dr^2 + 1/r\, du/dr = 1/\mu\, dp/dz \qquad (7\text{-}1)$$

$$u(R) = 0 \qquad (7\text{-}2)$$

$$du(0)/dr = 0,\ \text{or alternatively, "finite centerline 'u'"} \qquad (7\text{-}3)$$

Equation 7-1 is simply the axial momentum equation, while Equation 7-2 represents the "no-slip" velocity condition imposed at the solid wall r = R. We will comment on the centerline model in Equation 7-3 separately.

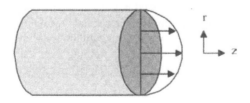

Figure 7-2. Flow in circular pipe.

In order to obtain the complete solution, we assume u(r) as the sum of "particular" and "complementary" solutions, that is, take $u(r) = u_p(r) + u_c(r)$ in the usual manner. To simplify the governing differential equation, we require that $u_p(r)$ satisfy $d^2u_p(r)/dr^2 + 1/r\, du_p/dr = 1/\mu\, dp/dz$. By inspection, we obtain

$u_p(r) = 1/(4\mu) \, dp/dz \, r^2$. Then, it is clear that $u_c(r)$ must satisfy the homogeneous ordinary differential equation, which has the solution "$u_c(r) = A + B \, \log_e r$" where A and B are constants whose values are to be determined. One constraint condition is Equation 7-2. The second is often expressed in two different ways. Usually, it is argued that "u" must be finite along the centerline $r = 0$; an alternative statement is the vanishing of viscous stress at the centerline, that is, $du(0)/dr = 0$, which also expresses a type of physical symmetry. In either case, $B = 0$. Simple manipulations then show that

$$u(r) = - 1/(4\mu) \, dp/dz \, (R^2 - r^2) \qquad (7\text{-}4)$$

$$Q = - \pi R^4 \, dp/dz \, /(8\mu) \qquad (7\text{-}5)$$

where Q is the constant volume flow rate through any pipe cross-sectional plane. It is important to observe the fourth-power dependence of Q on R, and the linear dependence on dp/dz, in Equation 7-5. We emphasize that these properties apply to Newtonian flows only.

Already, the limitations of the above "general" analysis are apparent. For example, consider again the arbitrary duct in Figure 7-1. Now, it may be that "no-slip" conditions can be applied, although not without numerical difficulty. However, more abstract questions arise. For example,

- "Where is the centerline?"

- If it is possible to define a 'centerline,' then, "what physical properties are satisfied along it?"

- "In the non-Newtonian case, is 'finite u' along the centerline sufficient to eliminate one integration constant?"

- "Since 'du(0)/dr = 0' does not apply to 'sharp, large n' axial velocity profiles, what is its generalization to nonlinear flow?"

These questions are difficult, and their answers are nontrivial. However, they only arise in the context of "radial" or "radial-like" coordinates. Later in this chapter, we will find that we need not answer them if we take the unusual step of reformulating pipe flow problems in special "rectangular-like," boundary conforming coordinates. In these variables, singular "$\log_e r$" type solutions never appear, and rightly so: physically, there is nothing ever "singular" within the flow domain away from the pipe walls, since pipe flows are generally smooth.

FINITE DIFFERENCE METHOD

Numerical methods provide powerful tools in the fluid-dynamicist's arsenal of mathematical tools. We will introduce the finite difference method here, to demonstrate an alternative solution procedure for the above problem; this general technique is used later in the solution of our grid generation and transformed flow equations.

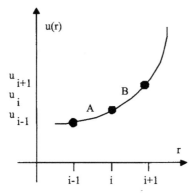

Figure 7-3. Finite difference indexing.

Finite differences are easily explained using ideas from calculus. The objective, in this section, requires our replacement of Equation 7-1 by an equivalent system of algebraic equations, which is easily solved by standard matrix methods. To accomplish this, we observe that the first derivatives at points "A" and "B" in Figure 7-3 are just $du(A)/dr = (u_i - u_{i-1})/\Delta r$ and $du(B)/dr = (u_{i+1} - u_i)/\Delta r$, where Δr is a constant mesh width, and "i" is an index used to mark successive divisions in r. Thus, the second derivative at "i" can be calculated as

$$d^2 u(r_i)/dr^2 = (du(B)/dr - du(A)/dr)/\Delta r$$
$$= (u_{i-1} - 2u_i + u_{i+1})/(\Delta r)^2 \qquad (7\text{-}6)$$

The first derivative at "i" appears more straightforwardly as

$$du(r_i)/dr = (u_{i+1} - u_{i-1})/(2\Delta r) \qquad (7\text{-}7)$$

If we now substitute Equations 7-6 and 7-7 in Equation 7-1, we obtain the resulting "central difference approximation"

$$(u_{i-1} - 2u_i + u_{i+1})/(\Delta r)^2 + (u_{i+1} - u_{i-1})/2r_i\Delta r = (1/\mu) \ dp/dz \quad (7\text{-}8)$$

which can be rearranged to give

$$\{1 - \Delta r/(2r_i)\}u_{i-1} + \{-2\}u_i + \{1 + \Delta r/(2r_i)\}u_{i+1} \qquad (7\text{-}9)$$
$$= (\Delta r)^2/\mu \ dp/dz$$

Equations 7-6 and 7-7 are "centrally" differenced because first and second derivatives at "i" involve values *both* to the left and right. It can be shown that central differences are more accurate than skewed "backward" or "forward" approximations; thus, for a fixed number of meshes, central differencing provides the greatest accuracy. In the language of numerical analysis, central difference methods are "second order" accurate.

In order to render our ideas more concretely, let us imagine that we approximate the domain $0 < r < R$ by "$i_{max} - 1$" number of grids, which is in turn bounded on the left and right by the indexes $i = 1$ and $i = i_{max}$. Then, if we write Equation 7-9 separately for $i = 2, 3, 4, ..., i_{max} - 1$, we obtain

$i = 2$: $\{1 - \Delta r / (2r_2)\}u_1 + \{-2\}u_2 + \{1 + \Delta r / (2r_2)\}u_3 = (\Delta r)^2/\mu \, dp/dz$

$i = 3$: $\{1 - \Delta r / (2r_3)\}u_2 + \{-2\}u_3 + \{1 + \Delta r / (2r_3)\}u_4 = (\Delta r)^2/\mu \, dp/dz$

$i = 4$: $\{1 - \Delta r / (2r_4)\}u_3 + \{-2\}u_4 + \{1 + \Delta r / (2r_4)\}u_5 = (\Delta r)^2/\mu \, dp/dz \, ...$

or,

$$\{\} \ u_1 + \{\} \ u_2 + \{\} \ u_3 \qquad\qquad = (\Delta r)^2/\mu \, dp/dz$$
$$\{\} \ u_2 + \{\} \ u_3 + \{\} \ u_4 \qquad = (\Delta r)^2/\mu \, dp/dz$$
$$\{\} \ u_3 + \{\} \ u_4 + \{\} \ u_5 = (\Delta r)^2/\mu \, dp/dz \, ...$$

If we now augment these equations with the "no-slip" condition in Equation 7-2, that is, $u_{imax} = 0$ at R, and the centerline condition of Equation 7-3, i.e., $u_1 - u_2 = 0$, we have "i_{max}" number of equations in "i_{max}" unknowns, namely,

$$u_1 - \quad u_2 \qquad\qquad\qquad\qquad = 0$$
$$\{\} \ u_1 + \{\} \ u_2 + \{\} \ u_3 \qquad\qquad = (\Delta r)^2/\mu \, dp/dz$$
$$\{\} \ u_2 + \{\} \ u_3 + \{\} \ u_4 \qquad = (\Delta r)^2/\mu \, dp/dz$$
$$\{\} \ u_3 + \{\} \ u_4 + \{\} \ u_5 = (\Delta r)^2/\mu \, dp/dz$$
$$\ldots\ldots$$
$$u_{imax} = 0$$

or, in matrix notation,

$$\{ \} \, u = R \qquad\qquad\qquad\qquad\qquad (7\text{-}10)$$

Note that the centerline $r = 0$ is located at $i = 1$; then, with $r_i = (i - 1)\Delta r$, the position $r = R$ at i_{max} requires us to choose $R = (i_{max} - 1)\Delta r$. The expanded form of Equation 7-10 involves three diagonals; for this reason, Equation 7-10 is known as a "tridiagonal" system of equations. Such equations are easily solved by widely available tridiagonal solvers. The solver is invoked once only, and numerical solutions for u_i or u are then available for $0 < r < R$.

In Figure 7-4, the Fortran code is given, which defines tridiagonal matrix coefficients based on Equation 7-9, for solution by the subroutine "TRIDI." A pressure gradient with $dp/dz = 0.001$ psi/ft is assumed, and a 1-inch radius is taken; also, we consider a unit centipoise viscosity fluid, with $\mu = 0.0000211$ lbf sec/ft^2. Units of "inch, sec, lbf" are used in the source listing. The program breaks the one-inch radial domain into ten equal increments, with $\Delta r = 0.1$ inch. Centerline symmetry conditions are imposed by Fortran statements at $i = 1$, while no-slip conditions are implemented at the wall $i = 11$. The "99's" are "dummies" used to "fill" unused matrix elements. This is necessary for certain computers, operating systems, and compilers, since undefined array elements will otherwise lead to error. Also programmed is the exact velocity solution in Equation 7-4, evaluated at its maximum value, along the centerline $r = 0$.

The printed solution below indicates that "Exact Umax = -.1422E+03 in/sec" while, on the basis of the finite difference model, "Speed = -.1391E+03 in/sec," thus incurring a small 2% error, in spite of the coarse mesh used.

```
I =  1, Speed = -.1391E+03 in/sec (Centerline, max velocity)
I =  2, Speed = -.1391E+03 in/sec
I =  3, Speed = -.1354E+03 in/sec
I =  4, Speed = -.1285E+03 in/sec
I =  5, Speed = -.1188E+03 in/sec
I =  6, Speed = -.1061E+03 in/sec
I =  7, Speed = -.9063E+02 in/sec
I =  8, Speed = -.7226E+02 in/sec
I =  9, Speed = -.5103E+02 in/sec
I = 10, Speed = -.2694E+02 in/sec
I = 11, Speed =  .0000E+00 in/sec (Wall, "no-slip" zero velocity)
      Exact Umax = -.1422E+03 in/sec
```

This exercise shows that ten grids blocks will suffice for computing most flows to engineering accuracy. On Pentium personal computers, this is accomplished in a "split second" calculation.

```
C       RADIAL.FOR
        DIMENSION AA(11),BB(11),CC(11),VV(11),WW(11)
        DELTAR = 0.1
        DPDZ = 0.001/12.
        VISC = 0.0000211/144.
        DO 100 I=2,10
        R = (I-1)*DELTAR
        AA(I) = +1. - DELTAR/(2.*R)
        BB(I) = -2.
        CC(I) = +1. + DELTAR/(2.*R)
        WW(I) =  (DELTAR**2.)*DPDZ/VISC
100     CONTINUE
        AA(1)  = 99.
        BB(1)  = 1.
        CC(1)  = -1.
        WW(1)  = 0.
        AA(11) = 0.
        BB(11) = 1.
        CC(11) = 99.
        WW(11) = 0.
        CALL TRIDI(AA,BB,CC,VV,WW,11)
        DO 200 I=1,11
        WRITE(*,130) I,VV(I)
130     FORMAT(1X,'I = ',I2,', Speed = ',E10.4,' in/sec')
200     CONTINUE
        WRITE(*,210)
210     FORMAT(' ')
        UMAX = - (DPDZ/(4.*VISC))*(1.0**2 - 0.**2)
        WRITE(*,230) UMAX
230     FORMAT('   Exact Umax = ',E10.4,' in/sec')
        STOP
        END
```

Figure 7-4a. Fortran code, Newtonian flow in circular pipe.

```
SUBROUTINE TRIDI(AA,BB,CC,VV,WW,N)
DIMENSION AA(11),BB(11),CC(11),VV(11),WW(11)
AA(N) = AA(N)/BB(N)
WW(N) = WW(N)/BB(N)
DO 100 I=2,N
II=-I+N+2
BN=1.0/(BB(II-1)-AA(II)*CC(II-1))
AA(II-1)=AA(II-1)*BN
WW(II-1)=(WW(II-1)-CC(II-1)*WW(II))*BN
100 CONTINUE
VV(1)=WW(1)
DO 200 I=2,N
VV(I)=WW(I)-AA(I)*VV(I-1)
200 CONTINUE
RETURN
END
```

Figure 7-4b. Fortran code, Newtonian flow in circular pipe (continued).

NEWTONIAN FLOW IN RECTANGULAR DUCTS

In this section, we provide a complementary exposition for Newtonian flow in rectangular ducts. The solutions obtained here and in the above analysis will be used to validate the general algorithm developed later for non-Newtonian flow in more general cross-sections. Here, we will observe that, even with our restriction to the simplest fluid, very different mathematical techniques are needed for a "simple" change in duct shape. From an engineering point of view, this is impractical: a more "robust" approach applicable to large classes of problems is needed, and motivated, particularly by the discussion given below.

Exact analytical series solution. Here, a closed form solution for unidirectional, laminar, steady, Newtonian viscous flow in a rectangular duct is obtained. Unlike Equation 7-1, which is an *ordinary* differential equation requiring additional data only at two separate points in space, we now have the *partial* differential equation

$$\partial^2 u/\partial x^2 + \partial^2 u/\partial y^2 = (1/\mu)\ dp/dz \tag{7-11}$$

Its solution is obtained, subject to the "no-slip" velocity boundary conditions

$$u(-\tfrac{1}{2}\,b < y < +\tfrac{1}{2}\,b,\ x = 0\,) = 0 \tag{7-12a}$$

$$u(-\tfrac{1}{2}\,b < y < +\tfrac{1}{2}\,b,\ x = c\,) = 0 \tag{7-12b}$$

$$u(y = -\tfrac{1}{2}\,b,\ 0 < x < c) = 0 \tag{7-12c}$$

$$u(y = +\tfrac{1}{2}\,b,\ 0 < x < c) = 0 \tag{7-12d}$$

where "b" and "c" denote the lengths of the sides of the rectangular duct shown in Figure 7-5.

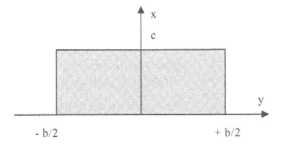

Figure 7-5. Rectangular duct, cross-section.

As before, the solution is obtained by taking $u(y,x)$ as the sum of "particular" and "complementary" solutions, that is, $u = u_p(x) + u_c(y,x)$. To simplify the analysis, we allow $u_p(x)$ to vanish at $x = 0$ and c, while satisfying $d^2u/dx^2 = (1/\mu)\, dp/dz$, where dp/dz is a prescribed constant. Then, the particular solution is obtained as $u_p(x) = dp/dz\, x^2/(2\mu) + C_1 x + C_2$, where the constants of integration can be evaluated to give $u_p(x) = -\,dp/dz\,(xc - x^2)/(2\mu)$. This involves no loss of generality since the complementary solution $u_c(y,x)$ has not yet been determined, and will be expressed as a function of $u_p(x)$. With this choice, the partial differential equation for $u_c(x)$ reduces to the classical Laplace equation

$$\partial^2 u_c/\partial x^2 + \partial^2 u_c/\partial y^2 = 0 \tag{7-13}$$

Now, since $u = 0$ and $u_p(x) = 0$ along the upper and lower edges of the rectangle in Figure 7-5, it follows that $u_c(x) = 0$ there also, since $u_c = u - u_p(x)$. By separating variables in the conventional manner, it is possible to show that product representations of $u_c(y,x)$ involve combinations of trigonometric and exponential functions. In particular, we are led to the combination

$$u_c = \sum_{n=1}^{\infty} A_n \cosh(n\pi y/c)\sin(n\pi x/c) \tag{7-14}$$

The factor "$\sin(n\pi x/c)$" allows $u_c(y,x)$ to vanish at the lower and upper boundaries $x = 0$ and c. The linear combination of exponentials "$\cosh(n\pi y/c)$" is selected because the velocity distribution must be symmetric with respect to the vertical line $y = 0$. Specific products cannot be disallowed, so the infinite summation accounts for the maximum number permitted. The coefficient A_n must be determined in such a way that side wall no-slip conditions are satisfied. To do this, we reconstruct the complete solution as

$$u = u_p(x) + u_c(y,x) = -\,dp/dz\,(xc - x^2)/(2\mu)$$
$$+ \sum A_n \cosh(n\pi y/c)\sin(n\pi x/c) \tag{7-15}$$

and apply the boundary conditions given by Equations 7-12a,b. The coefficients of the resulting Fourier series can be used, together with the orthogonality properties of the trigonometric sine function, to show that

$$A_n = dp/dz/(\mu c)\ c^3\ [\ 2 - \{2\cos(n\pi) + n\pi\sin(n\pi)\}/ \qquad (7\text{-}16)$$
$$[(n\pi)^3\cosh\{n\pi b/(2c)\}]$$

With A_n defined, the solution to u_c, and hence, to Equations 7-11 and 7-12, is determined. The shear rates $\partial u/\partial x$ and $\partial u/\partial y$, and viscous stresses $\mu\ \partial u/\partial x$ and $\mu\ \partial u/\partial y$, can be obtained by differentiating Equation 7-15. Again, analytical methods suffer limitations, e.g., the superpositions in "$u = u_p + u_c$" and "Σ" are not valid when the equation for "u" is nonlinear, as for non-Newtonian rheologies. Also, while there are no "log" function or "centerline" problems, as for radial formulations, it is clear that even if "y and x" coordinates are found for general ducts, it will not be possible to find the analogous "sin" and "cosh" functions. In general, for arbitrarily clogged ducts, there will be no lines of symmetry to help in defining solution products. Classical techniques are labor intensive in this sense: each problem requires its own special solution strategy. The Fortran code required to implement Equations 7-15 and 16 is shown in Figure 7-6. The input units will be explained later. Note that large values of the summation index "n" will lead to register overflow; thus, the apparent generality behind Equation 7-14 is limited by practical machine restrictions.

```
C       SERIES.FOR
C       INPUTS (Observation point (Y,X) assumed)
        B  = 1.
        C  = 1.
        Y  = 0.
        X  = 0.5
        VISC = 0.0000211/144.
        PGRAD = 0.001/12.
C       SOLUTION (Consider 100 terms in series)
        PI  = 3.14159
        C2  = C**2.
        SUM = 0.
        DO 100  N=1,100
        TEMP = 2.*(C**3)  - (C**3)*(2.*COS(N*PI) +N*PI*SIN(N*PI))
        TEMP = TEMP/((N**3.)*(PI**3.))
        A = PGRAD*TEMP/(VISC*C)
        A = A/COSH(N*PI*B/(2.*C))
        SUM = SUM + A*COSH(N*PI*Y/C)*SIN(N*PI*X/C)
  100   CONTINUE
        UC = SUM
        UP = -PGRAD*(C*X-X**2.)/(2.*VISC)
        U = UC + UP
        WRITE(*,200) U
  200   FORMAT(1X,'Velocity = ',E10.4,' in/sec')
        STOP
        END
```

Figure 7-6. Fortran code, series solution for rectangular duct.

Finite difference solution. Now, we obtain the solution for flow in a rectangular duct by purely numerical means. Like the analytical methods for circular and rectangular pipes, which are completely different, the same can be said of computational approaches. We emphasize that the solution for radial flow was obtained by "calling" the matrix solver just once. For problems in two independent variables, iterative methods are generally required to obtain practical solutions. For linear problems, say, Newtonian flows, it is possible to obtain the solution in a single pass using "direct solvers." However, these are not practical for complicated geometries, because numerous meshes are required to characterize the defining contours. In the analysis below, we will illustrate the use of iterative methods, since these are used in the solution of our governing grid generation and transformed flow equations.

We now turn to Equation 7-11 and consider it in its entirety, without resolving the dependent variable into particular and complementary parts. That is, we address $\partial^2 u/\partial x^2 + \partial^2 u/\partial y^2 = (1/\mu)\, dp/dz$ directly. From Equation 7-6, we had shown that

$$d^2 u(r_i)/dr^2 = (u_{i-1} - 2u_i + u_{i+1})/(\Delta r)^2 \qquad (7\text{-}17a)$$

Thus, we can similarly write

$$\partial^2 u(y_i)/\partial y^2 = (u_{i-1} - 2u_i + u_{i+1})/(\Delta y)^2 \qquad (7\text{-}17b)$$

for second derivatives in the "y" direction. In the present problem, we have an additional "x" direction, as shown in Figure 7-7. The grid depicted there overlays the cross-section of Figure 7-5. Since "y,x" requires two indexes, we extend Equation 7-17b in the obvious manner. For example, for a fixed j, the second derivative

$$\partial^2 u(y_i,x_j)/\partial y^2 = (u_{i-1\,j} - 2u_{i\,j} + u_{i+1\,j})/(\Delta y)^2 \qquad (7\text{-}17c)$$

Similarly,

$$\partial^2 u(y_i,x_j)/\partial x^2 = (u_{i\,j-1} - 2u_{i\,j} + u_{i\,j+1})/(\Delta x)^2 \qquad (7\text{-}17d)$$

Thus, at the "observation point" (i,j), Equation 7-11 becomes

$$\begin{aligned}(u_{i-1\,j} - 2u_{i\,j} + u_{i+1\,j})/(\Delta y)^2 & \\ + (u_{i\,j-1} - 2u_{i\,j} + u_{i\,j+1})/(\Delta x)^2 &= 1/\mu\, dp/dz\end{aligned} \qquad (7\text{-}18)$$

We can proceed to develop a rectangular duct solver allowing arbitrarily different Δx and Δy values. However, that is not our purpose. For simplicity, we will therefore assume constant meshes $\Delta x = \Delta y = \Delta$, which allows us to rewrite Equation 7-18 in the form

$$u_{i\,j} = \tfrac{1}{4}(u_{i-1\,j} + u_{i+1\,j} + u_{i\,j-1} + u_{i\,j+1}) - \Delta^2/(4\mu)\, dp/dz \qquad (7\text{-}19)$$

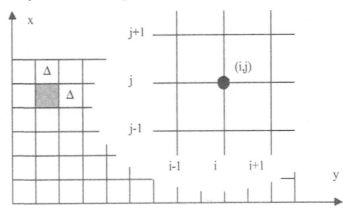

Figure 7-7. Rectangular finite difference grid.

Equation 7-19 is a central difference approximation to governing Equation 7-11, which is second-order accurate. Interestingly, it can be used as a "recursion formula" that iteratively produces improved numerical solutions. For example, suppose that some approximate solution for u(i,j) is available. Then, an improved (left side) solution can be generated by evaluating the right side of Equation 7-19 with it. It can be shown that, if this method converges, it will tend to the correct physical solution whatever the starting guess. Thus, if an initial approximation were not available, a trivial "zero solution" for u would be perfectly acceptable! Such methods are known as "relaxation methods." Since we have calculated improvements point-by-point (e.g., as opposed to an entire line of points at a time), the method used is a "point relaxation" method.

The Fortran source code implementing Equation 7-19 and the boundary conditions in Equations 7-12a,b,c,d is given in Figure 7-8. The units used are identical to those of the previous example, but here, a square duct having one-inch sides is considered. A mesh width of 0.1 inch is assumed, so that ten grids are taken along each side of the square. Loop 100 initializes the "starting guess" for U(I,J) to zero, also setting vanishing velocities along the duct walls I = 1 and 11, and J = 1 and 11. Loop 300 updates U(I,J) in the internal flow domain bounded by I = 2,..., 10 and J = 2, ..., 10. One hundred iterations are taken, which more than converges the calculation; in a more refined implementation, suitable convergence criteria would be defined. "Q" provides the volume flow rate in gallons per minute, while "U" is calculated in inches per second. For the Fortran code shown, computations are completed in less than one second, on Pentium class personal computers.

```
C       SQFDM.FOR (SQUARE DUCT, FINITE DIFFERENCE METHOD)
        DIMENSION U(11,11)
C       SQUARE IS 1" BY 1" AND THERE ARE 10 GRIDS
        DEL = 1./10.
        VISC = 0.0000211/144.
        PGRAD = 0.001/12.
C
        DO 100  I=1,11
        DO 100  J=1,11
        U(I,J) = 0.
100     CONTINUE
C
        DO 300  N=1,100
        DO 200  I=2,10
        DO 200  J=2,10
        U(I,J) =  (U(I-1,J) + U(I+1,J) + U(I,J-1) + U(I,J+1))/4.
     1        -   PGRAD*(DEL**2)/(4.*VISC)
200     CONTINUE
300     CONTINUE
        Q = 0.
        DO 400  I=2,11
        DO 400  J=2,11
        Q = Q + U(I-1,J-1)*(DEL**2)
400     CONTINUE
        Q = Q*0.2597
        Q = -Q
        WRITE(*,500) Q
500     FORMAT(' Volume flow rate = ',E10.4,' gal/min')
        WRITE(*,510) U(6,6)
510     FORMAT(' Umax = ',E10.4,' in/sec')
        STOP
        END
```

Figure 7-8. Finite difference code, rectangular ducts.

Example calculation. Here, a pressure gradient with dp/dz = 0.001 psi/ft is assumed, and a square duct with one-inch sides is taken; also, we consider a unit centipoise viscosity fluid, with μ = 0.0000211 lbf sec/ft^2. Units of "inch, sec, lbf" are used in the source listing. The program breaks each side of the square into ten equal increments, with $\Delta x = \Delta y = 0.1$ inch. This is done for comparative purposes with radial flow results. For the finite difference method, the maximum velocity is found at the center of the duct, that is, y = 0 and x = 0.5 inch, and it is given by the value u(6,6) = -0.4157E+02 in/sec. The code in Figure 7-6 gives the exact series solution at the center as -0.4190E+02, so that the difference method incurs less than 1% error. Again, this accuracy is achieved with a coarse "10×10" constant mesh.

GENERAL BOUNDARY CONFORMING GRID SYSTEMS

We have seen how analytical and numerical methods for circular and rectangular ducts require completely different solution strategies and techniques. Despite the sophistication of the approaches, conventional methods cannot be used to address more general duct shapes, e.g., flow passages in engines, pipes with clogs, "squashed" cross-sections, and so on, as typified by Figure 7-9.

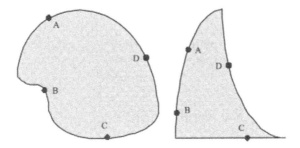

Figure 7-9. General pipe cross-section in (y,x) coordinates.

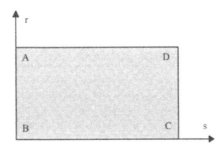

Figure 7-10. Pipe mapped to rectangular computational (r,s) space.

Without tools that can analyze flow efficiently and accurately, engineering questions of importance cannot be addressed. For instance, "What is the pressure drop required to 'start' a clogged pipe with a given initial volume flow rate?" "What is the shape that passes a prescribed flow rate with the least required pressure?" "How does this shape depend on fluid rheology?" "Are triangular cross-sections better than 'squashed' shapes?" Fortunately, the models developed here now permit perplexing questions such as these to be addressed. We will develop this computational technology from first principles.

In many engineering problems, a judicious choice of coordinate systems simplifies calculations and brings out the salient physical features more transparently than otherwise. For example, the use of cylindrical coordinates for single well problems in petroleum engineering leads to elegant "radial flow" results that are useful in well testing. Cartesian grids, on the other hand, are preferred in simulating oil and gas Darcy flows from rectangular fields.

However, despite the superficial similarities, there is significant mathematical difference between radial flow into a well and radial flow within a pipeline. In the former, the well (i.e., "centerline") does not form part of the computational domain, whereas in the latter, the centerline does. This leads to problems in more general duct geometries, where the notions of "centerline" and "centerline boundary conditions" are less clear. These questions have no obvious answer. The solution, then, is to abandon "radial like" formulations, and opt for "rectangular" grid systems instead. Importantly, "rectangular" systems can be used for non-rectangular shapes. For example, as suggested in Figures 7-9 and 7-10, general "circle-like" and "triangle-like" ducts can be topologically "mapped" into convenient rectangles, taking points A, B, C, and D arbitrarily as vertex points. We now review the new methodology.

We will draw upon results from differential geometry, which allow us to construct "boundary conforming, natural coordinates" for computation. These general techniques extend classical ideas on conformal mapping. They have accelerated progress in simulating aerospace flows past airfoils and cascades, and are only beginning to be applied in the petroleum industry. Thompson, Warsi and Mastin (1985) provides an excellent introduction to the subject.

To those familiar with conventional analysis, it may seem that the choice of (y,x) coordinates to solve problems for domains like those in Figure 7-9 is "unnatural." But our use of such coordinates was motivated by the new gridding methods that, like classical conformal mapping, are founded on Cartesian coordinates. We will give the "recipe" first, in order to make the approach more understandable. In the very first step, vertex points "A, B, C, and D" are defined for any given duct. Obviously, they should be judiciously spaced, e.g., not too closely, but far enough apart that each segment hosts some dominant geometric feature(s). Along each segment, that is, AB, BC, CD, and DA, sets of (y,x) points are selected, which form the outer "skeleton" of the grid to be defined.

The skeleton points chosen in the previous paragraph are now "transferred" to the edges of the rectangle in Figure 7-10. For example, if these points were (more or less) uniformly spaced in Figure 7-9, they can be uniformly spaced in Figure 7-10 for convenience. This is not a requirement; however, the "clockwise" or "directional" order of the selected points must be preserved in order to prevent overlapping curves. The next step in the approach requires us to solve a set of nonlinearly coupled, second-order PDEs, in particular,

$$(y_r^2 + x_r^2)\, y_{ss} - 2(y_s y_r + x_s x_r)\, y_{sr} + (y_s^2 + x_s^2)\, y_{rr} = 0 \qquad (2\text{-}36)$$

$$(y_r^2 + x_r^2)\, x_{ss} - 2(y_s y_r + x_s x_r)\, x_{sr} + (y_s^2 + x_s^2)\, x_{rr} = 0 \qquad (2\text{-}37)$$

The foregoing user-selected (y,x) values are applied as boundary conditions along the edges of the rectangle of Figure 7-10, and Equations 2-36, 27 are solved using a method not unlike the relaxation method developed earlier. Importantly, while the above boundary value problems are coupled and nonlinear, they *are* solved in a "rectangular computational space" that does not require imposition of boundary conditions along twisted spatial curves. In fact, (y,x) values are applied along lines of constant "r" and "s," thus allowing the use of widely available relaxation methods for Cartesian-based systems, such as the one for u(y,x) developed earlier; note that this "r" is not the radial coordinate variable used previously. Once the solutions for x(r,s) and y(r,s) are available for a given duct geometry, they are stored for future use.

We now turn our attention to the flow equations. While we had addressed Equation 7-11 above for simple Newtonian flow, we now redirect our thoughts to the most general non-Newtonian model for u(y,x) introduced in Chapter 2. We will first re-express this "u" through a different function u(r,s). For example, the simplified Equation 2-31 had transformed to

$$(y_r^2 + x_r^2)\, u_{ss} - 2(y_s y_r + x_s x_r)\, u_{sr}$$

$$+ (y_s^2 + x_s^2)\, u_{rr} = (y_s x_r - y_r x_s)^2\, \partial P/\partial z\, /N(\Gamma) \qquad (2\text{-}38)$$

whereas the result for the Equation 2-30 requires additional terms. For Equation 2-38 and its exact counterpart, the velocity terms in the apparent viscosity $N(\Gamma)$ of Equation 2-29 transform according to

$$u_y = (x_r u_s - x_s u_r)/(y_s x_r - y_r x_s) \qquad (2\text{-}39)$$

$$u_x = (y_s u_r - y_r u_s)/(y_s x_r - y_r x_s) \qquad (2\text{-}40)$$

The "u(r,s)" boundary value problem in Equation 2-38, subject to "no-slip" velocity boundary conditions along AB, BC, CD and DA in Figure 7-10, that is, lines of constant "r" and "s," is next solved. The variable coefficients in Equation 2-38 are evaluated from stored solutions for x(r,s) and y(r,s), which, we emphasize, are computed only once for a given duct shape.

As before, a "point relaxation" method, known formally as point "Successive Over Relaxation," or "SOR," can be used to solve Equation 2-38, although other variants of this iterative method can be used. These are known as "SLOR," representing "Successive Line Over Relaxation," which can be implemented in "row" or "column" form. These methods are discussed in standard numerical analysis references and will not be reviewed here. Once the solution for u(r,s) has converged, shear rates at a point in physical space can be computed from Equations 2-39 and 2-40. These partial derivatives can then be used to compute apparent viscosity and non-Newtonian viscous stress.

Recapitulation. We have replaced the differential equation for u(y,x) requiring boundary conditions along arbitrary curves, by two boundary value problems, namely, Equations 2-36 and 2-37, and Equation 2-38, which are solved in simple rectangular space. This added complication increases computing time, but the increase in accuracy offered by the method more than compensates for the additional resources required. In practice, though, complex duct flows can be computed in *less than one second* on Pentium personal computers, so that the increased level of computation is not a real impediment.

The mapping of "simple ducts" to rectangles is sketched In Figures 7-9 and 7-10. In Chapter 2, where the transformation method was first introduced, our annular spaces can be recognized as "ducts with single holes." In the language of mathematics, these are known as "doubly connected domains." They are complicated, but nonetheless amenable to similar solution approaches. The motivated reader should study that chapter, and note how our introduction of "branch cuts" effectively transformed doubly-connected to "singly connected" domains, not unlike those obtained for simple ducts. For flow domains with "multiple holes," additional sets of branch cuts can be introduced, to map any complicated region into duct-like domains. This subject is treated later.

We also emphasize that the final choice of grid used in any computation is dictated by numerical stability considerations. Different mesh patterns are associated with different convergence properties. For example, a "smooth" grid will produce converged results quickly, while a "kinky" or "rapidly varying" grid will lead to computational problems. In this sense, the exact choice of user selected skeleton points is important, and particularly so, along branch cuts.

Two example calculations. Given all the mathematical and programming complexities involved, benchmark calculations are desired to validate the basic computational engine. In the first simulation, we reconsider our one-inch-square duct, again assuming a 1 cp Newtonian fluid flowing under a 0.001 psi/ft pressure gradient. The curvilinear mesh program described above yields the following results,

```
Total volume flow rate  =  .5088E+01 gal/min
Umax = -.4166E+02 in/sec
Cross-sectional area  =  .1000E+01 sq inch
```

On the other hand, our earlier finite difference relaxation solution for the square duct, taking a square mesh in physical (y,x) space, gave

```
Volume flow rate  =  .5026E+01 gal/min
Umax = -.4157E+02 in/sec
```

Values for total flow rate and maximum centerline velocity agree to within 1%, thus providing more than enough accuracy for most engineering applications. The value of Umax from our exact analytical series solution was -.4190E+02 in/sec, which is almost identical to the curvilinear mesh result obtained above.

The fact that three completely different methods give identical solutions validates all of the approaches (and software code) developed in this research. Of course, the strength of the "boundary conforming mesh" approach is its ready extension to more complicated duct geometries. That different geometries can be compared using the same mathematical and software models means that relative comparisons are more reliable from an engineering point of view.

For example, the velocity plot shown in Figure 7-11 is obtained for the square duct considered above. (Color patterns do not show expected symmetries because different numbers of grids are taken horizontally and vertically; color values are determined by weighted averages in flattened gridblocks, whose aspect ratios differ at the top and the sides. This comment is directed toward square and circular ducts.) The coordinate points describing this duct were modified, as shown in Figure 7-12a to model debris, hydrate plugs, or wax icicles, in which case the figure is displayed upside-down. Note the complete "arbitrariness" of the shape assumed, and the ability of the algorithm to compute useful results. Using the same run parameters, we determine here that

```
Total volume flow rate  = .1476E+01 gal/min
Umax =  -.1473E+02 in/sec
Cross-sectional area = .7700E+00 sq inch
```

Compared with the unblocked square duct, a small 23% reduction in flow cross-sectional area leads to a *three-fold* reduction in volume flow rate*!* In general, flow rate reductions due to plugging will depend strongly on the shape of the plug, the "n, k" rheology of the fluid, and the applied pressure gradient.

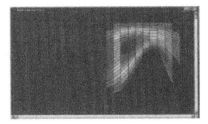

Figure 7-11. Velocity, clean square duct (see CDROM for color slides).

Figure 7-12a. Velocity, clogged square duct with large plug (see CDROM for color slides).

Figure 7-12b. Stress "N(Γ) $\partial u/\partial x$."

Figure 7-12d. Dissipation function (see CDROM for color slides).

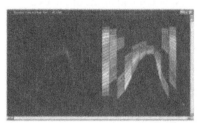

Figure 7-12c. Stress "N(Γ) $\partial u/\partial y$."

Total flow rate provides one "snapshot" of a plugged duct. In practice, we wish to study its "stability" also, that is, the plug's likelihood to erode or increase in size. This tendency can be inferred from the distribution of mechanical viscous stress acting along the plug surfaces. Figures 7-12b,c, for example, show how "N(Γ) $\partial u/\partial x$" and "N(Γ) $\partial u/\partial y$" behave quite differently at different areas of the plug. The dissipation function, introduced in Chapter 2, is a measure of heat generated by internal fluid friction, which is likely to be small. However, because it involves both rectangular components of viscous stress, it is also a good indicator of total stress and therefore "erodability at a point."

Figure 7-12d suggests that the upper left side of the plug is likely to erode. Low net stresses at the bottom right (together with low velocities, as observed from Figure 7-12a) suggest that the bottom right portion of the duct will worsen with time. In the next chapter, we will deal with pipe plugging in greater detail, and discuss various issues associated with debris accumulation, wax deposition, and hydrate formation in straight circular pipes containing non-Newtonian flow. Note that the duct simulations described in this chapter have direct application to heating and air conditioning system design.

CLOGGED ANNULUS AND STUCK PIPE MODELING

The heavily clogged "annuli" shown in Figures 7-13a,b, with black debris occupying the bottom, arise in stuck pipe analysis in horizontal drilling, and possibly, in start-up modeling for bundled pipelines when fluid in the conduit has gelled. Depending on the texture of the interface, e.g., loose sands, cohesive wax particles, semi-gelled mud, the erosion or cleaning mechanism involved may correlate with velocity, viscous shear stress, or combinations of the two.

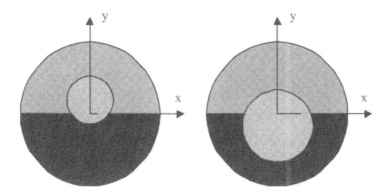

Figure 7-13a,b. Heavily clogged annuli, start-up conditions.

The exact physical mechanisms are not apparent, and remain the subject of industry research. Laboratory rheometers measure "intrinsic" properties like "n," "k," and yield stress, and should not be used to correlate with test or field scale observations pertaining to cuttings transport effectiveness or material removal rate. These lab measurements are physically related to "global" properties like shear rate, viscous stress, and apparent viscosity, which are functions of the test fixture under real world flow conditions. These global properties can only be obtained by computational simulation since there is little practical chance that dimensionless scale-up can be applied.

We emphasize that the "annuli" shown in Figures 7-13a,b are not true geometric annuli, since the marked gray zones do not contain "holes." These gray zones are ducts, at least topologically; the four vertices used in the mapping transform are the obvious "corners" located at the gray-black interface. In Figures 7-14a to 7-14f, we have computed various properties associated with a power law fluid (see CDROM for color slides). In particular, Figure 7-14b shows how apparent viscosity varies throughout the cross-section; also, we note that the dissipation function, which measures internal heat generation, is also an indicator of total stress.

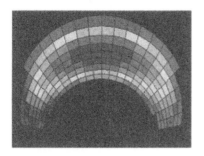

Figure 7-14a. Axial velocity.

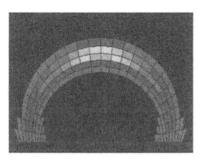

Figure 7-14b. Apparent viscosity.

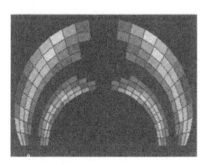

Figure 7-14c. Viscous stress,
"N(Γ) $\partial u/\partial x$."

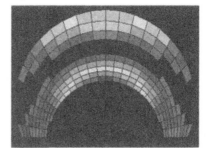

Figure 7-14d. Viscous stress,
"N(Γ) $\partial u/\partial y$."

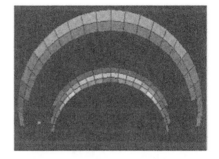

Figure 7-14e. Dissipation
function.

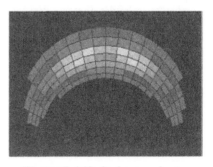

Figure 7-14f. Stokes product.

Let us now repeat the above calculations for a Bingham plastic, and, in particular, simulate a "plug flow" for the foregoing "annulus." Solutions for Bingham plastics in pipes and annuli are well known, but plug flow domains for less-than-ideal conduits have not been published. Here, we describe typical results and examine the quality of the solution. Figure 7-15a displays the calculated velocity distribution; a large yield stress was selected to obtain the expansive plug shown. Figures 7-15b,c for both rectangular components of the shear rate are consistent with plug behavior, since computed values are constant over the same area. The dissipation function in Figure 7-15d, measuring total stress, is similarly consistent. These qualities lend credibility to the algorithm. Color slides of these figures are provided in the CDROM.

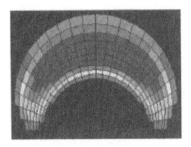

Figure 7-15a. Axial velocity.

Figure 7-15c. Shear rate $\partial u/\partial y$.

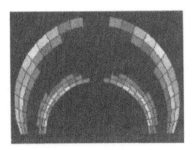

Figure 7-15b. Shear rate $\partial u/\partial x$.

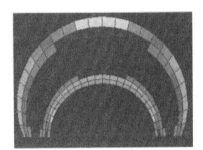

Figure 7-15d. Dissipation function.

Finally, let us consider the impending clog of an annular flow previously considered in Chapter 6. In Figure 7-16, the high velocity that would normally obtain at the bottom (wide part of the annulus) is interestingly displaced to the left and right sides as the conduit fills up with debris. Thus, various "interesting" flow patterns are possible, on the way to very clogged configurations such as the ones in Figures 7-13 and 7-14.

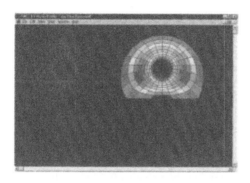

Figure 7-16. Velocity, for an impending annular clog.

(See CDROM for color slides.)

REFERENCES

Tamamidis, P., and Assanis, D.N., "Generation of Orthogonal Grids with Control of Spacing," *Journal of Computational Physics*, Vol. 94, 1991, pp. 437-453.

Thompson, J.F., "Grid Generation Techniques in Computational Fluid Dynamics," *AIAA Journal*, November 1984, pp. 1505-1523.

Thompson, J.F., Warsi, Z.U.A., and Mastin, C.W., *Numerical Grid Generation*, Elsevier Science Publishing, New York, 1985.

8

Solids Deposition Modeling

What is "solids deposition modeling" and what is its role in pipe or annular flow dynamics? Although numerous studies have been directed, for instance, at wax deposition and hydrate formation, none have addressed the dynamic interaction between the solids deposition process and the velocity field imparted by the flowing non-Newtonian fluid. The latter serves dual functions: it assists with solid particle placement, but at the same time, the viscous stress field associated with it tends to remove particles that have adhered to solid surfaces.

Until now, determining the velocity field alone has proven difficult, if not impossible: nonlinear flow equations must be solved for geometric domains that are far from ideal in shape. However, the methods developed in Chapter 2 for annular flow and Chapter 7 for general duct flow permit fast and robust solutions, and also, efficient post-processing and visual display for quantities like apparent viscosity, shear rate, and viscous stress. In this sense, "half" of the problem has been resolved, and in this chapter, we address the remaining half.

In order to understand the overall philosophy, it is useful to return to the problem of mudcake formation and erosion, and of cuttings transport, first considered in Chapter 5. As we have noted, the plugging or cleaning of a borehole annulus can be a dynamic, time-dependent process. For example, the inability of the low side flow to remove cuttings results in debris bed formation, when cuttings combine with mudcake to form mechanical structures. Forced filtration of drilling mud into the formation compacts these beds, and individual particle identities are lost: the resulting beds, characterized by well-defined yield stresses, alter the shape of the borehole annulus and the properties of the flow.

But the bed can be eroded or removed, provided the viscous stress imparted by the flowing mud in the modified annulus exceeds the yield value. If this is not possible, plugging will result and stuck pipe is possible. On the other hand, alternative remedial actions are possible. The driller can change the composition of the mud to promote more effective cleaning, increase the volumetric flow rate, or both. Successfully doing so erodes cuttings accumulations, and ideally, promotes dynamic "self-cleaning" of the hole.

In a sense, developing a new "constitutive relation," e.g., postulating Newtonian or power law properties, and deriving complementary flow equations is simpler than designing solids deposition models. The mathematical process needed to "place" stress-strain relations in momentum differential equation form is more straightforward than the cognitive process required to understand every step of a new physical phenomenon, e.g., wax deposition or hydrate formation. In this chapter, we introduce a philosophy behind modeling solids deposition, and as a first step, develop a simple model for mudcake and cuttings bed buildup over porous rock. We emphasize that there are no simple answers: each problem is unique, and the developmental process is very iterative.

Mudcake buildup on porous rock. Borehole annuli are lined with slowly thickening mudcake that, over large time scales, will reduce cross-sectional size. However, dynamic equilibrium is usually achieved because erosive forces in the flow stream limit such thickening. As a first step in understanding this process, growth in the absence of erosion must be characterized, but even this requires a detailed picture of the physics. The reader should carefully consider the steps needed in designing deposition models, taking this example as a model.

Since the permeability of the formation greatly exceeds that of mudcake, and the thickness of mudcake is small compared with the borehole radius, we can model cake growth in the idealized lineal flow test setup in Figure 8-1. We consider a one-dimensional experiment where mud, in essence a suspension of clay particles in water, is allowed to flow through filter paper. Initially, the flow rate is rapid. But as time progresses, solid particles (typically 6%-40% by volume for light to heavy muds) such as barite are deposited onto the surface of the paper, forming a mudcake that, in turn, retards the passage of mud filtrate by virtue of the resistance to flow that the cake provides.

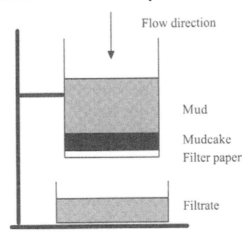

Figure 8-1. Simple laboratory mudcake buildup experiment.

We therefore expect filtrate volume flow rate and cake growth rate to decrease with time, while filtrate volume and cake thickness continue to increase, but ever more slowly. These qualitative ideas can be formulated precisely because the problem *is* based on well-defined physical processes. For one, the composition of the homogeneous mud during this filtration does not change: its *solid fraction* is always constant. Secondly, the flow *within* the mudcake *is* a Darcy flow, and is therefore governed by the equations used by reservoir engineers. The only problem, though, is the presence of a *moving boundary*, namely, the position interface separating the mudcake from the mud that ultimately passes through it, and which continually adds to its thickness. The physical problem, therefore, is a *transient* process that requires somewhat different mathematics than that taught in partial differential equations courses.

Mudcakes in reality may be compressible, that is, their mechanical properties may vary with applied pressure differential. We will be able to draw upon reservoir engineering methods developed for subsidence and formation compaction later. For now, a simple constitutive model for incompressible mudcake buildup, that is, the filtration of a fluid suspension of solid particles by a porous but rigid mudcake, can be constructed from first principles. First, let $x_c(t) > 0$ represent cake thickness as a function of the time, where $x_c = 0$ indicates zero initial thickness. Also, let V_s and V_l denote the volumes of solids and liquids in the mud suspension, and let f_s denote the *solid fraction* defined by $f_s = V_s/(V_s + V_l)$. Since this does not change throughout the filtration, its time derivative must vanish. If we set $df_s/dt = (V_s + V_l)^{-1} dV_s/dt - V_s (V_s + V_l)^{-2} (dV_s/dt + dV_l/dt) = 0$, we can show that $dV_s = (V_s/V_l) dV_l$. But since, separately, $V_s/V_l = f_s/(1- f_s)$, it follows that $dV_s = \{f_s/(1- f_s)\} dV_l$. This is, essentially, a *conservation of species* law for the solid particles making up the mud suspension, and does not as yet embody any assumptions related to mudcake buildup. Frequently, we might note, the drilling fluid is thickened or thinned in the process of making hole; if so, the equations derived here should be reworked with $f_s = f_s(t)$ and its corresponding time-dependent pressure drop.

In order to introduce the mudcake dynamics, we observe that the total volume of solids dV_s deposited on an elemental area dA of filter paper during an infinitesimal time dt is $dV_s = (1 - \phi_c) dA\, dx_c$ where ϕ_c is the mudcake porosity. During this time, the volume of filtrate flowing through our filter paper screen is $dV_l = |v_n|\, dA\, dt$ where $|v_n|$ is the Darcy velocity of the filtrate through the cake and past the paper. We now set our two expressions for dV_s equal, in order to form $\{f_s/(1- f_s)\} dV_l = (1 - \phi_c) dA\, dx_c$, and replace dV_l with $|v_n|\, dA\, dt$, so that we obtain $\{f_s/(1- f_s)\} |v_n|\, dA\, dt = (1 - \phi_c) dA\, dx_c$.

Now, it is seen that the dA's cancel, and we are led to a generic equation governing mudcake growth. In particular, the cake thickness $x_c(t)$ satisfies the ordinary differential equation

$$dx_c(t)/dt = \{f_s/\{(1-f_s)(1-\phi_c)\}\} |v_n| \qquad (8\text{-}1a)$$

At this point, we assume a one-dimensional, constant density, single *liquid* flow. For such flows, the constant Darcy velocity is $(k/\mu)(\Delta p/L)$, where $\Delta p > 0$ is the usual "delta p" pressure drop through the core of length L, assuming that a Newtonian approximation applies. The corresponding velocity for the present problem is $|v_n| = (k/\mu)(\Delta p/x_c)$ where k is the mudcake permeability, and μ is a mean filtrate viscosity. Substitution in Equation 8-1a leads to

$$dx_c(t)/dt = \{kf_s\Delta p/\{\mu(1-f_s)(1-\phi_c)\}\}/x_c \qquad (8\text{-}1b)$$

If the mudcake thickness is infinitesimally thin at t = 0, with $x_c(0) = 0$, Equation 8-1b can be integrated, with the result that

$$x_c(t) = \sqrt{[\{2kf_s\Delta p/\{\mu(1-f_s)(1-\phi_c)\}\} t]} > 0 \qquad (8\text{-}1c)$$

This demonstrates that cake thickness in a lineal flow grows with time like $\sqrt{t}$. However, it grows ever more slowly, because increasing thickness means increasing resistance to filtrate through-flow, the source of the solid particulates required for mudcake buildup; consequently, filtrate buildup also slows.

To obtain the filtrate production volume, we combine $dV_l = |v_n|$ dA dt and $|v_n| = (k/\mu)(\Delta p/x_c)$ to form $dV_l = (k\Delta p dA/\mu) x_c^{-1}dt$. Using Equation 8-1c, we find $dV_l = (k\Delta p dA/\mu)[\{2kf_s\Delta p/\{\mu(1-f_s)(1-\phi_c)\}\}]^{-1/2}(t)^{-1/2}$ dt. Direct integration, assuming zero filtrate initially, yields

$$V_l(t) = 2(k\Delta p dA/\mu) [\{2kf_s\Delta p/\{\mu(1-f_s)(1-\phi_c)\}\}]^{-1/2}(t)^{1/2} \qquad (8\text{-}1d)$$
$$= \sqrt{\{2k\Delta p(1-f_s)(1-\phi_c)/(\mu f_s)\}} \sqrt{t} \ dA$$

This correctly reproduces the common observation that filtrate volume increases in time like "$\sqrt{t}$." The mudcake deposition model in Equation 8-1c, at this point, is credible, and is significant in that it explicitly highlights the roles of the individual parameters k, f_s, Δp, μ, and ϕ_c.

Now, along the walls of general boreholes that are not necessarily circular, containing drillpipes that need not be concentric, the "$x_c(t)$" in Equation 8-1c would apply at each location; of course, "$x_c(t)$" must be measured in a direction perpendicular to the local surface area. This thickness increases with time by the same amount everywhere; consequently, the hole area decreases and the annular geometry changes, with more pronounced curvature. At the same time, drilling fluid is flowing parallel to the borehole axis. This flow, generally non-Newtonian, must be calculated using the methods developed in Chapters 2 and 7. The mechanical yield stress τ_y of the formed cake, which must be separately determined in the laboratory, is an important physical constant of the system. If the stress τ imparted by the fluid is less than τ_y, a very simple deposition model might allow Equation 8-1c to proceed "as is." However, if $\tau > \tau_y$ applies locally, one might postulate that, instead of Equation 8-1c, an "erosion model"

$$dx_c(t)/dt = f(\,\ldots) \tag{8-2}$$

where the function "f < 0" might depend on net flow rate, gel level, weighting material characteristics, and the magnitude of the difference "$\tau - \tau_y$." In unconsolidated sands penetrated by deviated wells, "f" may vary azimuthally, since gravity effects at the top of the hole differ from those at the bottom. And in highly eccentric annuli, mudcake at the low side may be thicker than high side cake, because lower viscous stress levels are less effective in cake removal.

Again, the mudcake buildup and removal process is time-dependent, and very dynamic, at least computationally. In the present example, we conceptually initialize calculations with a given eccentric annulus, possibly contaminated by cake, and calculate the non-Newtonian flow characteristics associated with this initial state. Equations 8-1c and 8-2 are applied at the next time step, to determine modifications to the initial shape. Then, flow calculations are repeated, with the entire process continuing until some clear indicator of hole equilibrium is achieved. The hole may tend to plug, in which case remedial planning is suggested, or it may tend to remain open.

In any event, the development of deposition and erosion models such as those in Equations 8-1c and 8-2 requires a detailed understanding of the physics, and consequently, calls for supporting laboratory experiments. As this example for mudcake deposition shows, it *is* possible to formulate phenomenological models analytically when the "pieces of the puzzle" are well understood, as we have for the "$\sqrt{t}$" model governing mudcake growth.

By the same token, it should be clear that in other areas of solids deposition modeling, for example, accumulation of produced fines, wax buildup, and hydrate plug formation in pipelines, "simple answers" are not yet available. More than likely, the particular models used will depend on the reservoir in question, and will probably change throughout the life of the reservoir. For this reason, the present chapter focuses on generic questions, and attempts to build a sound research approach and modeling philosophy for workers entering the field. At the present time, much of the published research on wax deposition and hydrate formation focuses on fundamental processes like crystal growth and thermodynamics. An experimental database providing even qualitative information is not yet available for detailed model development. Nonetheless, we can speculate on how typical models may appear, and comment on the mathematical forms in which they can be expressed.

DEPOSITION MECHANICS

In this section, we introduce the reader to basic ideas in different areas of solids deposition and transport by fluid flow, if only to highlight common physical processes and mathematical methods. By far, the most comprehensive literature is found in sedimentary transport and slurry movement, specialties that are well developed in civil engineering over decades of research. The following survey articles provide an excellent introduction to established techniques:

- Anderson, A.G., "Sedimentation," *Handbook of Fluid Mechanics*, V.L. Streeter, Editor, McGraw-Hill, New York, 1961.

- Kapfer, W.H., "Flow of Sludges and Slurries," *Piping Handbook*, R. King, Editor, McGraw-Hill, New York, 1973.

These references, in fact, motivated the cuttings transport research in Chapter 5. Concepts and results from these and related works are covered next.

Sedimentary transport. Sediment transport is important to river, shoreline, and harbor projects. The distinction between "cohesive" and "noncohesive" sediments is usually made. For example, clays are cohesive, while sand and gravel in stream beds consist of discrete particles. In cohesive sediments, the resistance to erosion depends primarily on the strength of the cohesive bond between the particles. Variables affecting particle lift-off include parameters like bed shear stress, fluid viscosity, and particle size, shape, and mass density, and number density distribution. Different forces are involved in holding grains down and entraining them into the flow. These include gravity, frictional resistance along grain contacts, "cohesiveness" or "stickiness" of clays due to electrochemical attraction, and forces parallel to the bed such as shear stress. The "sediment transporting capacity" of a moving fluid is the maximum rate at which moving fluid can transport a particular sediment aggregation.

Lift forces are perpendicular to the flow direction, and depend on the shapes of individual particles. For example, a stationary spherical grain in a uniform stream experiences no lift, since upper and lower flowfields are symmetric; however, a spinning or "tumbling" spherical grain will. On the other hand, flat grains oriented at nonzero angles with respect to the uniform flow do experience lift, whose existence is apparent from asymmetry. Of course, oncoming flows need not be uniform. It turns out that small heavy particles that have settled in a lighter viscous fluid can resuspend if the mixture is exposed to a shear field. This interaction between gravity and shear-induced fluxes strongly depends on particle size and shape. Note that the above force differs from the lift for airplane wings: small grains "see" low Reynolds number flows, while much larger bodies operate at high Reynolds numbers. Thus, formulas obtained in different fluid specialties must be carefully evaluated before they are used in deposition modeling. In either case, mathematical analysis is very difficult.

Once lifted into the flow stream, overall movement is dictated by the vertical "settling velocity" of the particle, and the velocity in the main flow. Settling velocity is determined by balancing buoyancy and laminar drag forces, with the latter strongly dependent on fluid rheology. For Newtonian flows, the classic Stokes solution applies; for non-Newtonian flows, analytical solutions are not available. Different motions are possible. Finer silts and clays will more or less float within a moving fluid. On the other hand, sand and gravel are likely to travel close to bed; those that "roll and drag" along the bottom move by traction, while those that "hop, skip, and jump" move by the process of saltation.

In general, modeling non-Newtonian flow past single stationary particles represents a difficult endeavor, even for the most accomplished mathematicians. Flows past unconstrained bodies are even more challenging. Finally, modeling flows past aggregates of particles is likely to be impossible, without additional simplifying statistical assumptions. For these reasons, useful and practical deposition and transport models are likely to be empirical, so that scalable laboratory experiments are highly encouraged. Simpler "ideal" flow setups that shed physical insight on key parameters are likely to be more useful than "practical, engineering" examples that include too many interacting variables.

Slurry transport. A large body of literature exists for slurry transport, e.g., coal slurries, slurries in mining applications, slurries in process plants, and so on. A comprehensive review is neither possible nor necessary, since water is the carrier fluid in the majority of references. But many fundamental ideas and approaches apply. Early references provide discussions on sewage sludge removal, emphasizing prevalent non-Newtonian behavior, while acknowledging that computations are not practical. They also discuss settling phenomena in slurries, e.g., the influence of particle size, particle density, and fluid viscosity.

"Minimum velocity" formulas are available that, under the assumptions cited, are useful in ensuring clean ducts when the carrier fluid is water. The notion of "critical tractive force," i.e., the value of shear stress at which bed movement initiates, is introduced; this concept was important in our discussions of cutting transport. Both "velocity" and "stress" criteria are used later in this chapter to construct illustrative numerical models of eroding flows. Also, the distinction between transport in closed conduits and open channels is made.

The literature additionally addresses the effects of channel obstructions and the formation of sediment waves; again, restrictions to water as the carrier fluid, are required. Numerous empirical formulas for Q that would give clean conduits are available in the literature; however, their applicability to oilfield debris, waxes, and hydrates is uncertain. While we carefully distinguish between velocity and stress as distinctly different erosion mechanisms, we note that, in some flow, the distinction is less clear. At times, for example, the decrease in bed shear stress is primarily a function of decreasing flow velocity.

WAXES AND PARAFFINS, BASIC IDEAS

As hot crude flows from reservoirs into cold pipelines, with low temperatures typical under deep subsea conditions, wax crystals may form along solid surfaces when wall temperatures drop below the "cloud point" or "wax appearance temperature." Crystals may grow in size until the wall is fully covered, with the possibility of encapsulating oil in the wax layers. Wax deposition can grow preferentially on one side of the pipe due to gravity segregation. As wax thickness builds, the pressure drop along the pipe must be increased to maintain constant flow rate, and power requirements increase. Constant pressure processes would yield decreasing flow rates.

Paraffin deposition can be controlled through various means. Insulation and direct heating pipe will reduce exposure to the cold environment. Mechanical pigging is possible. Chemical inhibitors are also used. For example, surfactants or dispersants alter the ability of wax particles to adhere to each other or to pipe wall surfaces; in the language of sedimentary transport, they become less cohesive, and behave more like discrete entities. Biochemical methods, for instance, use of bacteria to control wax growth.

In this book, we will address the effect of nonlinear fluid rheology and noncircular duct flow in facilitating wax erosion. The "critical tractive force" ideas developed in slurry transport, extended in Chapter 5 to cuttings removal, again apply to bed-like deposits. Recent authors, for example, introduce "critical wax tension" analogously, defined as the critical shear force required to remove a unit thickness of wax deposit; the exact magnitude depends on oil composition, wax content, temperature, buildup history, and aging.

More complications. Paraffin deposition involves thermodynamics, but other operational consequences arise that draw from all physical disciplines.

- Electrokinetic effects may be important with heavy organic constituents. Potential differences along the conduit may develop due to the motion of charged particles; these induce alterations in colloidal particle charges downstream which promote deposition. That is, electrical charges in the crude may encourage migration of separated waxes to the pipe wall.

- In low flow rate pipelines, certain waxes sink because of gravity, and form sludge layers at the low side. Also, density segregation can also lead to recirculating flows of the type modeled in Chapter 4.

- For lighter waxes, buoyancy can cause precipitated wax to collect at the top of the pipe (in the simulations performed in this chapter, no distinction is made between "top" and "bottom," since our "snapshots" can be turned "upside-down").

- Deposited wax will increase wall roughness and therefore increase friction, thus reducing pipeline flow capacity.

- Suspended particulates such as asphaltenes, formation fines, corrosion products, silt, and sand, for instance, may encourage wax precipitation, acting as nuclei for wax separation and accumulation. Wax particles so separated may not necessarily deposit along walls; they may remain in suspension, altering the rheology of the carrier fluid, affecting its ability to "throw" particles against pipe walls or to remove wax deposits by erosion.

- Although significant deposition is unlikely under isothermal conditions, that is, when pipeline crude and ocean temperatures are in equilibrium, wall deposits may nonetheless form. Pipe roughness, for instance, can initiate stacking, leading to local accumulations that may further grow.

Wax precipitation in detail. Waxy crude may contain a variety of light and intermediate hydrocarbons, e.g., paraffins, aromatics, naphtenic, wax, heavy organic compounds, and low amount of resins, asphaltenes, organo-metallics. Wax in crudes consists of paraffin (C18-C36) and naphtenic (C30-C60) hydrocarbons. These wax components exist in various states, that is, gas, liquid, or solid, depending on temperature and pressure. When wax freezes, crystals are formed. Those formed from paraffin wax are known as "macrocrystalline wax," while those originating from naphtenes are "microcrystalline."

When the temperature of a waxy crude is decreased, the heavier fractions in wax content appear first. The "cloud point" or "wax appearance temperature" is the temperature below which the oil is saturated with wax. Deposition occurs when the temperature of the crude falls below cloud point. Paraffin will precipitate under slight changes in equilibrium conditions, causing loss of solubility of the wax in the crude. Wax nucleation and growth may occur along the pipe surface and within the bulk fluid. Precipitation within the fluid causes its viscosity to increase and alters the non-Newtonian characteristics of the carrier fluid. Increases in frictional drag may initiate pumping problems and higher overall pipe pressures. Note that the carrier fluid is rarely a single-phase flow. More often than not, wax deposition occurs in three-phase oil, water, and gas flow, over a range of gas-oil ratios, water cuts, and flow patterns, which can vary significantly with pipe inclination angle.

Wax deposition control. The most direct means of control, though not necessarily the least inexpensive, target wall temperature by insulation or heating, possibly through internally heated pipes as discussed in Chapter 6. But the environment is far from certain. Some deposits do not disappear on heating and are not fully removed by pigging. Crudes may contain heavy organics like asphaltene and resin, which may not crystallize upon cooling and may not have definite freezing points; these interact with wax differently, and may prevent wax crystal formation or enhance it. Solvents provide a different alternative. However, those containing benzene, ethyl benzene, toluene, and so on, are encountering opposition from regulatory and environmental concerns. The problems are acute for offshore applications; inexpensive and environmentally friendly control approaches with minimal operational impact are desired.

Wax growth on solid surfaces, under static conditions, is believed to occur by molecular diffusion. Behind most deposition descriptions are liquid phase models and equations of state, with the exact composition of the wax phase determined by the model and the physical properties of the petroleum fractions. We do not attempt to understand the detailed processes behind wax precipitation and deposition in this book. Instead, we focus on fluid-dynamical modeling issues, demonstrating how non-Newtonian flows can be calculated for difficult "real world" duct geometries that are less than ideal. The "mere" determination of the flowfield itself is significant, since it provides information to evaluate different modes of deposition and to address important remediation issues.

For example, as in sedimentary transport, flow nonuniformities play dual roles: they may "throw" particles onto surfaces, where they adhere, or they can remove buildups by viscous shear. Both effects must be studied, in light of experimental data, using the background velocity and stress fields our analysis provides. The modeling approaches reported in this book hope to establish the hydrodynamic backbone that makes accurate modeling of these phenomena possible. For example, is it possible to design a fluid that keeps particles suspended, or perhaps, to understand the conditions under which the flow is self-cleaning? What are rheological effects of chemical solvents? Wax can cause crude oil to gel and deposit on tubular surfaces. What shear stresses are required to remove them? And finally, waxy crude oil may gel after a period of shutdown. What levels of pressure are required to initiate start-up of flow?

Modeling dynamic wax deposition. In principle, modeling the dynamic, time-dependent interaction between waxy deposits attempting to grow, and duct flows attempting to erode them, is similar to, although slightly more complicated than, the mudcake model developed earlier. The *deposition*, or *growth* model, shown conceptually in Figure 8-2a, consists of two parts, namely, a thermal component in which buildup is driven by temperature gradients, and a mechanical component in which velocity "throws" additional particles that have precipitated in the bulk fluid into the wax-lined pipe surface. This velocity may be coupled to the temperature environment, as discussed in Chapter 6. Various solids convection models are available in the fluids literature, and, in general, different deposition models are needed in different production scenarios.

The competing *erosive* model is schematically shown in Figure 8-2b, in which we emphasize the role of non-Newtonian fluid stress at the walls; it is similar to our model for cuttings transport removal from stiff beds. Wax yield stress may be determined in the laboratory or inferred from mechanical pigging data, e.g., see Souza Mendes *et al.* (1999) or related pipeline literature.

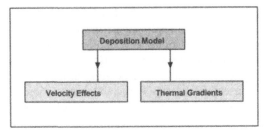

Figure 8-2a. Conceptual deposition model.

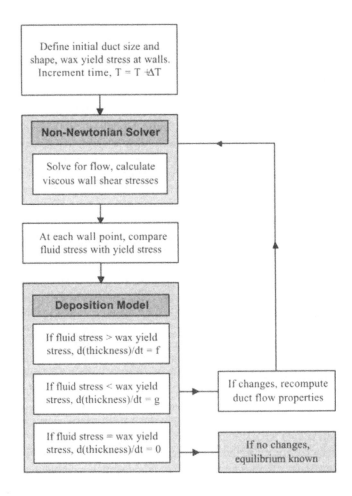

Figure 8-2b. Fluid flow and solids deposition model interaction.

Hydrate control. Natural gas production from deep waters can be operationally hampered by pipeline plugging due to gas hydrates. Predicting the effects of pipe hydraulics on hydrate behavior is necessary to achieving optimal hydrate control. As exploration moves offshore, the need to minimize production facility construction and maintenance costs becomes important. Producers are seeking options that permit the transport of unprocessed fluids miles from wellheads or subsea production templates to central processing facilities located in shallower water. Deep-water, multiphase flow lines can offer cost saving benefits to operators and, consequently, basic and applied research related to hydrate control is an active area of interest.

Hydrate crystallization takes place when natural gas and water come into contact at low temperature and high pressure. Hydrates are "ice-like" solids that form when sufficient amounts of water are available, a "hydrate former" is present, and the proper combinations of temperatures and pressures are conducive. Gas hydrates are crystalline compounds that form whenever water contacts the constituents found in natural gas, gas condensates, and oils, at the hydrate formation equilibrium temperatures and pressures, as Figure 8-3 shows. Hydrate crystals can be thought of as integrated networks of hydrogen-bonded, "soccer ball"-shaped ice cages with gas constituents trapped within.

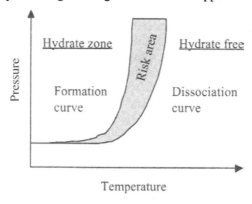

Figure 8-3. Hydrate dependence on "P" and "T."

Low seabed temperatures and high pressures can significantly impact the commercial risk of deepwater projects. Hydrates can cause plugging, an unacceptable condition, given the inaccessibility of deep subsea pipelines. Hydrate plugging is not new, and early on, profoundly affected onshore production and flow. But these problems became less severe as hydrate phase equilibrium data became available; these data provided the basis for modern engineering and chemical inhibition procedures using methanol and glycol. Such treatments can be costly in deep water, though, given the quantities of inhibitor required, not to mention expensive storage facilities; but these approaches remain attractive, as recent research has led the way to more effective, low toxicity compounds as useful alternatives to methanol or glycol. Field and laboratory studies have had some success, but problems remain that must be solved before the industry gains advantages in utilizing these inhibitors.

Operational considerations are also important to hydrate mitigation. Proper amounts of chemicals must arrive at target flowline locations at the required time to control the rate of crystal formation, growth, agglomeration, and deposition. This combined chemical and hydrodynamic control strategy in general multiphase pipeline environments must be effective over extended shut-in periods to accommodate a range of potential offshore operating scenarios.

Understanding the effects of chemicals on rheology and flow represents one aspect of the mitigation problem. In pipeline plugging, we are concerned, as noted above, with the effects of obstructions on pressure drops and flow rates. On the other hand, natural hydrates represent a potentially important source of natural gas, although they can potentially clog pipelines. One possible delivery solution is to convert associated gases into frozen hydrates, which are then mixed with refrigerated crude oil, to form slurries, which are in turn pumped through pipelines and into shuttle tankers for transport to shore. By blending ground hydrates with suitable carrier fluids, a transportable slurry can be formed that efficiently delivers "gas" to market.

Several questions are immediately apparent. How finely should hydrates be ground? What is the ideal "solids in fluid" concentration? Fineness, of course, influences rheology; the solids that remain affect plugging, and the combination controls delivery economics. And what happens as hydrates convect into higher pressure pipeline regimes? In any event, we are concerned with the pumpability of the slurry, and also, the ability of the slurry to erode hydrate plugs that have formed in the flow path. These considerations require a model that is able to simulate flows in duct geometries that are far from circular. With it, we can simulate worst case conditions and optimize operations.

In this chapter, we will not focus on the physics and chemistry of hydrate formation, the kinetics of formation and agglomeration, or the physiochemical characterization of the solid constituents. Instead, we will study flows past "hydrate plugs." Wax buildup is "predictable" to the extent that depositions can be found at top and low sides, and all too often, azimuthally about the entire circumference. Hydrates, in contrast, may appear "randomly." For example, they can form as layers separating gas on the top side and water on the low side. In terms of size, hydrate particles may vary from finely dispersed solid particles to big lumps that stick to the walls of pipelines. Hydrate particle size is nonuniform and follows wide distribution densities. But, in general, large plugs can be found almost anywhere, a situation that challenges non-Newtonian flow modeling in arbitrary ducts. Simulation is important in defining start-up procedures, because large plugs are associated with extremely large pressure drops that may be difficult to achieve in practice.

Pipe inclination may play a significant role for denser fluids. Ibraheem *et al* (1998) observe that, for their horizontal and 45° positions, predictions may be optimistic since lift forces, virtual mass effects, and so on, are not incorporated, and that a two-dimensional model will be necessary. This caution is well justified. In Chapter 4, we showed that density stratification can lead to recirculation vortices that plug the pipeline, while in Chapter 5, we showed that 45°-70° inclinations are worst, even when density variations are ignored.

Recapitulation. Very subtle questions are possible. Can hydrate pipeline blockages lead to increased flowline pressures that facilitate additional hydrate growth? Can viscous shear stresses developed within a carrier fluid, or perhaps a hydrate slurry, that support "self-cleaning," which in turn eliminates isolated plugs that form? Again, the formalism developed in Figure 8-2b for wax removal applies, but now with Figure 8-2a replaced by one applicable to hydrate formation. We will show that numerical simulations can be conveniently performed for large, asymmetrically shaped plugs, that is, our grid generation and velocity solvers are truly "robust" in the numerical sense. Thus, it is clear that the simulation methodology also applies to other types of conduits, valves, and fittings that can potentially support hydrate formation.

MODELING CONCEPTS AND INTEGRATION

Our mathematical description of time-dependent mudcake buildup, without erosive effects, is relevant to wax buildup under nonisothermal conditions. Recall that once cake starts building, incremental growth of cake retards further buildup, since additional resistance impedes fluid filtration. Thus, the rate of cake growth should vary inversely with cake thickness; in fact, we had shown

$$dx_C(t)/dt = \{kf_s\Delta p/\{\mu(1- f_s)(1 -\phi_C)\}\}/x_C \tag{8-1b}$$

Direct integration of "$x_C\,dx_C = ..$" leads "½ $x^2 = ..t$," that is, the "√t law,"

$$x_C(t) = \sqrt{[\{2kf_s\Delta p/\{\mu(1- f_s)(1 -\phi_C)\}\}\ t]} > 0 \tag{8-1c}$$

In this section, we introduce some elementary, but preliminary ideas, with the hope of stimulating further research. These following illustrative examples were designed to be simple, to show how mathematics and physics go hand in hand.

Wax buildup due to temperature differences. Paraphrasing the above, "once wax starts building, incremental growth of wax retards further buildup, since additional insulation impedes heat transfer." Let R_{pipe} denote the inner radius of the pipe, which is constant, and let $R(t) < R_{pipe}$ denote the time-varying radius of the wax-to-fluid interface. In cake buildup, growth rate is proportional to the pressure gradient; here, it is proportional to the heat transfer rate, or temperature gradient $(T -T_{pipe})/(R - R_{pipe})$ by virtue of Fourier's law of conduction, with T being the fluid temperature. We therefore write, analogously to Equation 8-1b,

$$dR/dt = \alpha \, (T - T_{pipe})/(R - R_{pipe}) \qquad (8\text{-}3)$$

where $\alpha > 0$ is an empirically determined constant. Cross-multiplying leads to $(R - R_{pipe}) \, dR = \alpha \, (T - T_{pipe}) \, dt$ where $T - T_{pipe} > 0$. Direct integration yields

$$\tfrac{1}{2} \, (R - R_{pipe})^2 = \alpha \, (T - T_{pipe}) \, t > 0 \qquad (8\text{-}4)$$

where we have used the initial condition $R(0) = R_{pipe}$ when $t = 0$.

Hence, according to this simple model, the thickness of the wax will vary as $\sqrt{t}$ under static conditions. Of course, in reality, α may depend weakly on T, crystalline structure, and other factors, and deviations from "$\sqrt{t}$" behavior are not unexpected. Furthermore, it is not completely clear that Equation 8-3 in its present form is correct; for example, dR/dt might be replaced by dR^n/dt, but in any event, guidance from experimental data is necessary. This buildup model treats wax deposition due to thermal gradients, but obviously, other modes exist. For general problems in arbitrarily shaped ducts, wax particles, debris, and fines convected with the fluid may impinge against pipe walls at rates proportional to local velocity gradients; or, they may deposit at low or high sides by way of gravity segregation, either because they are heavy or they are buoyant.

Simulating erosion. Again, any model is necessarily motivated by empirical observation, so our arguments are only plausible. For non-Newtonian flow in circular pipes, it is generally true that

$$\tau(r) = r \, \Delta p/2L > 0 \qquad (1\text{-}2a)$$

$$\tau_w = R \, \Delta p/2L > 0 \qquad (1\text{-}2b)$$

These equations are interesting because they show how shear stress τ must decrease as R decreases: thus, any wax buildup must be accompanied by lower levels of stress, and hence, decreases in the ability to self-clean or erode the wax. The most simplistic erosion model might take the form

$$dR/dt = \beta \, (\tau - \tau_y) > 0 \qquad (8\text{-}5)$$

where $\beta > 0$ is an empirical constant, $\tau - \tau_y > 0$, and τ_y is the yield stress of the wax coating. Thus, R increases with time, i.e., the cross-section "opens up." The uncertainties again remain, e.g., R can be replaced by R^2. Note that Equations 1-2a and 1-2b do not apply to annular flows.

Deposition and flowfield interaction. Our solution of the nonlinear rheology equations on curvilinear meshes is "straightforward" because the problem is at least well defined and tractable numerically. But the same cannot be said for wax or hydrate deposition modeling, since each individual application must be treated on a customized basis. As we have suggested in the above discussions, numerous variables enter, even in the simplest problems. For example, these include particle size, shape and distribution, cohesiveness, buoyancy, heat transfer, multiphase fluid flow, dissolved wax type, debris content, fluid rheology, pipeline characteristics, surface roughness, insulation, centrifugal force due to bends, volume flow rate, and so on.

Nonetheless, when a particular engineering problem is well understood, the dominant interactions can be identified, and integrated fluid flow and wax or hydrate deposition models can be constructed. The following simulations demonstrate different types of integrated models that have been designed to simulate flows in clogging and self-cleaning pipelines. These examples illustrate the broad range of applications that are possible, where the computational "engines" developed in Chapters 2 and 7 have proven invaluable in simulating operational reality.

DETAILED CALCULATED EXAMPLES

In this section, six simulation examples are discussed in detail. These demonstrate how the general duct model can be used to host different types of solids deposition mechanisms. However, the exact "constitutive relations" used are proprietary to the funding companies and cannot be listed here.

Simulation 1. Wax Deposition with Newtonian Flow in Circular Duct

In this first simulation set, we consider a unit centipoise Newtonian fluid, flowing in an initially circular duct; in particular, we assume a 6-inch radius, so that the cross-sectional area is 113.1 square inches. A family of "smile-shaped" surfaces is selected for the solids buildup boundary family of curves, since wax surfaces are expected to be more curved than flat. This buildup increases with time, and for convenience, the final duct cross-section is assumed to be an exact semi-circle, whose area is 113.1/2 or 56.55 square inches. A deposition model is invoked, and intermediate "cross-sectional area versus volume flow rate" results, assuming an axial pressure gradient of 0.001 psi/ft, at selected time intervals, are given in Figure 8-4 below.

Area (in^2)	Rate (gpm)	
.1129E+03	.7503E+05	(full circle)
.1082E+03	.6931E+05	
.1035E+03	.6266E+05	
.9882E+02	.5670E+05	
.9411E+02	.5090E+05	
.8941E+02	.4531E+05	
.8470E+02	.3994E+05	
.8000E+02	.3483E+05	
.7529E+02	.3000E+05	
.7059E+02	.2549E+05	
.6588E+02	.2132E+05	
.6117E+02	.1752E+05	
.5647E+02	.1411E+05	(semi-circle)

Figure 8-4. Flow rate versus duct area, with dp/dz fixed.

How do we know that computed results are accurate? We selected Newtonian flow for this validation because the Hagen-Poiseuille volume flow rate formula (e.g., see Chapter 1) for *circular* pipes can be used to check our numbers. This classic solution, assuming dp/dz = 0.001 psi/ft, R = 6 inches, and μ = 1 cp, shows that the flow rate is exactly .755E+05, as compared to our .750E+05 gpm. The ratio 755/750 is 1.007, thus yielding 0.7% accuracy.

Another indicator of accuracy is found in our computation of area. Obviously, the formula "πR^2" applies to our starting shape, which again yields 113.1 square inches. However, we have indicated 112.9 in Figure 8-4, for a 0.2% error. Why an error at all? This appears because our general topological analysis never utilizes "πR^2." The formulation is expressed in terms of metrics of the transformations x(r,s) and y(r,s). Therefore, if computed circle areas agree with "πR^2" and volume flow rates are consistent with classical Hagen-Poiseuille flow results, our mathematical boundary value problems, numerical analysis, and programming are likely to be correct. The last entry in Figure 8-4 gives our area for the semi-circle, which is to be compared with an exact 113.1/2 or 56.55 square inches. From the ratio 56.55/56.47 = 1.001, our "error" of 0.1% suggests that the accompanying .1411E+05 gpm rate is also likely to be correct.

Interestingly, from the top and bottom lines of Figure 8-4, it is seen that *a 50 percent reduction in flow area, from "fully circular" to "semi-circular," is responsible for a five-fold decrease in volume throughput.* This demonstrates the severe consequence of even partial blockage. Because the flow is Newtonian and linear in this example, the italicized conclusion is "scalable" and applicable to all Newtonian flows. That is, it applies to pipes of all radii R, to all pressure gradients dp/dz, and to all viscosities μ.

Why is "scalability" a property of Newtonian flows? To see that this is true, we return to the governing equation "$(\partial^2/\partial x^2 + \partial^2/\partial y^2)$ u(x,y) = 1/μ dp/dz" in the duct coordinates (x,y). Suppose that a solution u(x,y) for a given value of the "1/μ dp/dz" is available. If we replace this by "C/μ dp/dz," where C is a constant, it is clear that Cu must solve the modified problem. Similarly, if Q and τ represent total volume flow rate and shear stress in the original problem, the corresponding rescaled values are CQ and Cτ This would not be true if, for example, if μ were a nonlinear function of ∂u/∂x and ∂u/∂y, as in the case of non-Newtonian fluids; and if it were, it is now obvious that μ, or "N(Γ)," in the notation of Chapter 2, it must now vary with x and y because Γ depends on ∂u/∂x and ∂u/∂y. Interestingly, we have deduced these important properties even without "solving" the differential equation!

Unfortunately, in the case of non-Newtonian fluids, generalizations such as these cannot be made, and each problem must be considered individually. The extrapolations available to linear mathematical analysis are just not available. It is instructive to examine in detail, the velocity, apparent viscosity, shear rate, viscous shear stress distributions, and so on, for the similar sequence of simulations for non-Newtonian flows. Because generalizations cannot be offered, we do not need to quote the exact parameters assumed. Figures 8-4a to 8-4h provide "time lapse" results for a power law fluid simulation; note, for example, how apparent viscosities are not constant, but, in fact, vary throughout the cross-sectional area of the duct.

Our methodology and software allow us to plot all quantities of physical interest at each time step. Again, these quantities are needed to interpret solids deposition data obtained in research flow loop experiments, because deposition mechanisms are not very well understood. Due to space limitations, only the first and last "snapshots," plus an intermediate one, are shown; in the final time step, our initially circular duct has become purely semi-circular. The varied "snapshots" shown are also instructive because, to the author's knowledge, similar detailed results have never before appeared in the literature. Note that the enclosed CDROM includes a comprehensive set of 12 "time-lapse" color slides per physical property, detailing the complete evolution of the plugging.

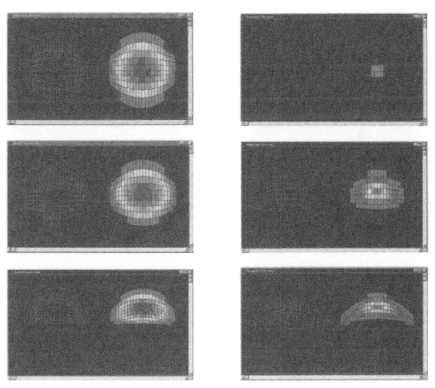

Figure 8-4a. Time Lapse Sequence: Axial Velocity "U."

Figure 8-4b. Time Lapse Sequence: Apparent Viscosity "N(Γ)."

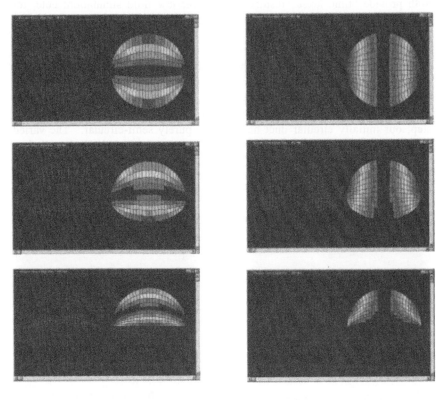

Figure 8-4c. Time Lapse
Sequence: Viscous Stress
"N(Γ) ∂U/∂x."

Figure 8-4d. Time Lapse
Sequence: Viscous Stress
"N(Γ) ∂U/∂y."

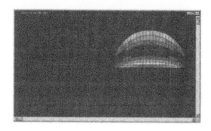

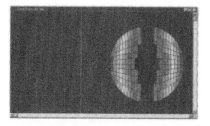

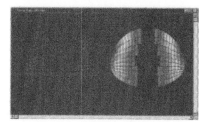

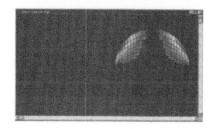

Figure 8-4e. Time Lapse
Sequence: Shear Rate "$\partial U/\partial x$."

Figure 8-4f. Time Lapse
Sequence: Shear Rate "$\partial U/\partial y$."

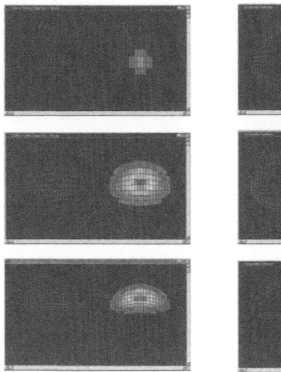

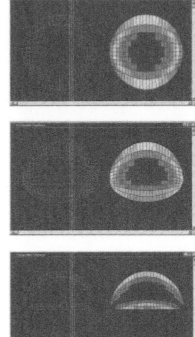

Figure 8-4g. Time Lapse
Sequence: Stokes' Product
"N(Γ)U."

Figure 8-4h. Time Lapse
Sequence: Dissipation Function
"Φ."

Simulation 2. Hydrate Plug with Newtonian Flow in Circular Duct
(Velocity Field)

In this second simulation, we consider the flow about an isolated, but growing "hydrate plug." This model does not offer any geometric symmetry, because, in reality, hydrate blockages can form "randomly" within the duct cross-section. Thus, our curvilinear grid algorithms are particularly useful in modeling real flowfields, and determining pressure drops associated with plugs having different shapes. For now, we again assume Newtonian flow so that our results are "scalable" in the sense of the previous example. This is *not* a limitation of the solver, which handles very nonlinear, non-Newtonian fluids. A Newtonian flow is assumed here only to provide results that can be generalized dimensionlessly and therefore are of greater utility to the reader, e.g., refer to the italicized conclusion in the earlier example.

In order to demonstrate the wealth of physical quantities that can be produced by the simulator, we have duplicated typical "high level" summaries; detailed area distributions of all quantities are, of course, available. Note that the assumed pressure gradient of "1 psi/ft" was taken for convenience only, and leads to flow rates that are somewhat large. However, because the flow is Newtonian, a thousand-fold reduction in pressure gradient will lead to a thousand-fold decrease in flow rate. Shear rates and viscous stresses scale similarly. This ability to rescale results makes our tabulated quantities useful in obtaining preliminary engineering estimates. In the following pages, example results of six time steps are selected for display. Detailed numerical results, for example, showing "typical" shear rates and viscous stresses, whose magnitudes must be rescaled in accordance with the above paragraph, are given first. Then, "snapshots" of the axial velocity field are given, in the same time sequence.

First run, initial full circle, without hydrate plug:

```
NEWTONIAN FLOW OPTION SELECTED.
Newtonian flow, constant viscosity = 1.00000 cp
Axial pressure gradient assumed as 1.0000E+01 psi/ft.
Total volume flow rate  = .7503E+08 gal/min
Cross-sectional area = .1129E+03 sq inch

TABULATION OF CALCULATED AVERAGE QUANTITIES:
Area weighted means for absolute values taken
over entire pipe (x,y) cross-sectional area
O  Axial flow velocity = .2266E+07 in/sec
O  Apparent viscosity = .1465E-06 lbf sec/sq in
O  Viscous stress, AppVis x dU/dx, = .1029E+00 psi
O  Viscous stress, AppVis x dU/dy, = .1230E+00 psi
O  Dissipation function = .2415E+06 lbf/(sec sq in)
O  Shear rate dU/dx = .7022E+06 1/sec
O  Shear rate dU/dy = .8394E+06 1/sec
O  Stokes product = .3321E+00 lbf/in
```

Second run:

```
NEWTONIAN FLOW OPTION SELECTED.
Newtonian flow, constant viscosity = 1.00000 cp
Axial pressure gradient assumed as .1000E+01 psi/ft.
Total volume flow rate  = .6925E+08 gal/min
Cross-sectional area = .1088E+03 sq inch
TABULATION OF CALCULATED AVERAGE QUANTITIES:
Area weighted means for absolute values taken
over entire pipe (x,y) cross-sectional area
O  Axial flow velocity = .2159E+07 in/sec
O  Apparent viscosity = .1465E-06 lbf sec/sq in
O  Viscous stress, AppVis x dU/dx, = .1050E+00 psi
O  Viscous stress, AppVis x dU/dy, = .1176E+00 psi
O  Dissipation function = .2350E+06 lbf/(sec sq in)
O  Shear rate dU/dx = .7168E+06 1/sec
O  Shear rate dU/dy = .8026E+06 1/sec
O  Stokes product = .3163E+00 lbf/in
```

Third run:

```
NEWTONIAN FLOW OPTION SELECTED.
Newtonian flow, constant viscosity = 1.00000 cp
Axial pressure gradient assumed as .1000E+01 psi/ft.
Total volume flow rate   = .6032E+08 gal/min
Cross-sectional area = .1047E+03 sq inch
```

```
TABULATION OF CALCULATED AVERAGE QUANTITIES:
Area weighted means for absolute values taken
over entire pipe (x,y) cross-sectional area
O  Axial flow velocity = .1974E+07 in/sec
O  Apparent viscosity = .1465E-06 lbf sec/sq in
O  Viscous stress, AppVis x dU/dx, = .1021E+00 psi
O  Viscous stress, AppVis x dU/dy, = .1066E+00 psi
O  Dissipation function = .2102E+06 lbf/(sec sq in)
O  Shear rate dU/dx = .6969E+06 1/sec
O  Shear rate dU/dy = .7275E+06 1/sec
O  Stokes product = .2893E+00 lbf/in
```

Fourth run:

```
NEWTONIAN FLOW OPTION SELECTED.
Newtonian flow, constant viscosity = 1.00000 cp
Axial pressure gradient assumed as .1000E+01 psi/ft.
Total volume flow rate   = .4253E+08 gal/min
Cross-sectional area = .9642E+02 sq inch
```

```
TABULATION OF CALCULATED AVERAGE QUANTITIES:
Area weighted means for absolute values taken
over entire pipe (x,y) cross-sectional area
O  Axial flow velocity = .1538E+07 in/sec
O  Apparent viscosity = .1465E-06 lbf sec/sq in
O  Viscous stress, AppVis x dU/dx, = .9147E-01 psi
O  Viscous stress, AppVis x dU/dy, = .8822E-01 psi
O  Dissipation function = .1638E+06 lbf/(sec sq in)
O  Shear rate dU/dx = .6243E+06 1/sec
O  Shear rate dU/dy = .6021E+06 1/sec
O  Stokes product = .2254E+00 lbf/in
```

Fifth run:

```
NEWTONIAN FLOW OPTION SELECTED.
Newtonian flow, constant viscosity = 1.00000 cp
Axial pressure gradient assumed as .1000E+01 psi/ft.
Total volume flow rate   = .3417E+08 gal/min
Cross-sectional area = .9229E+02 sq inch
```

```
TABULATION OF CALCULATED AVERAGE QUANTITIES:
Area weighted means for absolute values taken
over entire pipe (x,y) cross-sectional area
O  Axial flow velocity = .1300E+07 in/sec
O  Apparent viscosity = .1465E-06 lbf sec/sq in
O  Viscous stress, AppVis x dU/dx, = .8285E-01 psi
O  Viscous stress, AppVis x dU/dy, = .7919E-01 psi
O  Dissipation function = .1363E+06 lbf/(sec sq in)
O  Shear rate dU/dx = .5654E+06 1/sec
O  Shear rate dU/dy = .5405E+06 1/sec
O  Stokes product = .1905E+00 lbf/in
```

Sixth, final run, with large blockage:

```
NEWTONIAN FLOW OPTION SELECTED.
Newtonian flow, constant viscosity = 1.00000 cp
Axial pressure gradient assumed as .1000E+01 psi/ft.
Total volume flow rate  = .2711E+08 gal/min
Cross-sectional area = .8816E+02 sq inch

TABULATION OF CALCULATED AVERAGE QUANTITIES:
Area weighted means for absolute values taken
over entire pipe (x,y) cross-sectional area
O  Axial flow velocity = .1070E+07 in/sec
O  Apparent viscosity = .1465E-06 lbf sec/sq in
O  Viscous stress, AppVis x dU/dx, = .7323E-01 psi
O  Viscous stress, AppVis x dU/dy, = .7136E-01 psi
O  Dissipation function = .1115E+06 lbf/(sec sq in)
O  Shear rate dU/dx = .4997E+06 1/sec
O  Shear rate dU/dy = .4870E+06 1/sec
O  Stokes product = .1568E+00 lbf/in
```

In Figures 8-5a to 8-5f, sequential "snapshots" of the axial velocity field associated with a growing plug are shown. Color slides of these figures are available on the enclosed CDROM. The reader should refer to the foregoing listings for the corresponding duct areas, volume flow rates, average shear rates and stresses, and so on. How is "scalability" applied? Consider, for example, that "1 psi/ft" implies a shear rate component of ".4997E+06 1/sec" in the last printout. A more practical "0.001 psi/ft" would be associated with a shear rate of ".4997E+03 1/sec."

It is also interesting to compare the first and final runs. Initially, the full circle has an area of 112.9 square inches, and a volume flow rate of .7503E+08 gpm. In the last simulation, these numbers reduce to 88.16 and .2711E+08. Thus, a 22% reduction in flow area is responsible for a 64% decrease in flow rate! It is clear that even "minor" flowline blockages are not tolerable. Following these velocity diagrams, some discussion of the stress fields associated with the worst case blockage is given.

Simulation 3. Hydrate Plug with Newtonian Flow in Circular Duct (Viscous Stress Field)

In this example, we continue with Simulation 2 above, but focus on the largest blockage obtained in the final "snapshot." In particular, we consider the likelihood that the plug-like structure will remain in the form shown, given the erosive environment imparted by viscous shear stresses. To facilitate our discussion, we refer to Figure 8-6, which defines boundary points A, B, C, D and E, and also, interior point F. Figure 8-7a displays the "Stokes product," proportional to the product of apparent viscosity and velocity, which measures how well individual particles are convected with the flow. The maximum is located at F, where "in stream" debris are likely to be found.

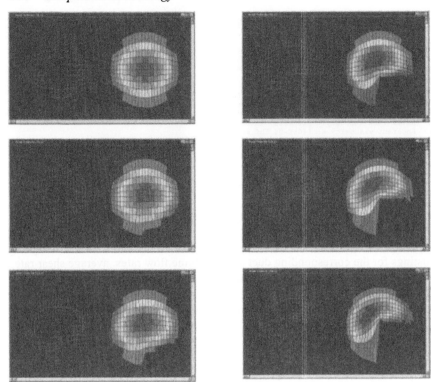

Figures 8-5a,b,c. Velocity field, hydrate plug formation.

Figures 8-5d,e,f. Velocity field, hydrate plug formation.

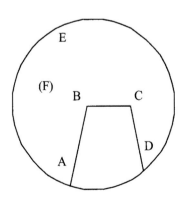

Figure 8-6. Generic plug diagram.

Figures 8-7b and 8-7c display both rectangular components of the viscous stress. The stresses $N(\Gamma)\ \partial u/\partial x$ and $N(\Gamma)\ \partial u/\partial y$ are strong, respectively, along BC and AB. Figure 8-7d shows the spatial distribution of the "dissipation function," which measures local heat generation due to internal friction, likely to be insignificant. However, the same function is also an indicator of total stress, which acts to erode surfaces that can yield. This figure suggests that "B" is most likely to erode. At the same time, stresses about our "hydrate plug" are lowest at "D," suggesting that additional local growth is possible. Color slides of these figures are available on the enclosed CDROM.

Figure 8-7a. Stokes product.

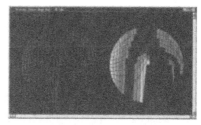

Figure 8-7c. Viscous stress, $N(\Gamma)$ $\partial u/\partial y$.

Figure 8-7b. Viscous stress, $N(\Gamma)$ $\partial u/\partial x$.

Figure 8-7d. Dissipation function.

Simulation 4. Hydrate Plug with Power Law Flow in Circular Duct

---

In this example, we study the flow of a non-Newtonian power law fluid past the worst case blockage in Simulation 3. In particular, we examine the "total volume flow rate versus axial pressure gradient," or "Q vs. dp/dz" signature of the flow. Before proceeding, it is instructive to reconsider the *exact* solution for power law flow in a circular pipe, namely,

$$Q/(\pi R^3) = [R\Delta p/(2kL)]^{1/n} n/(3n+1) \qquad (1\text{-}4c)$$

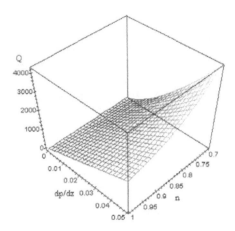

Figure 8-8. "Q vs. dp/dz" for various "n."

Results for "Q versus dp/dz" are plotted in Figure 8-8 for different values of "n," assuming a 6-inch-radius pipe and a fixed 'k" value that would correspond to 100,000 cp if n = 1. In the Newtonian flow limit of n = 1, linearity is clearly seen, however, this exact solution shows that pronounced curvature is obtained as "n" decreases from unity. For any fixed value of dp/dz, it is also seen that Q is strongly dependent on the power law index.

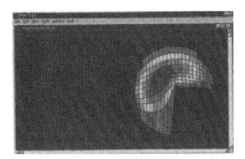

Figure 8-9. Typical power law velocity profile (see CDROM).

We are interested in the corresponding results for power law flow past the large blockage in the previous simulation. A number of runs were performed, holding fluid properties and geometry fixed, while "dp/dz" was varied. The particular values were selected because they gave "practical" volume flow rates. When dp/dz = 0.01 psi/ft, a flow rate of 651 gpm is obtained; at 0.10 psi/ft, the volume flow rate is not "6,510" but 11,570 gpm, clearly demonstrating the effects of nonlinearity. Values for dp/dz are shown in bold font, in the tabulated results reproduced below, and "Q versus dp/dz" is plotted in Figure 8-10.

First run:

```
POWER LAW FLOW OPTION SELECTED.
Power law fluid assumed, with exponent "n" equal
to .8000E+00 and consistency factor of .1000E-03
lbf sec^n/sq in.
```

```
Axial pressure gradient assumed as .1000E-01 psi/ft.
Total volume flow rate   = .6508E+03 gal/min
Cross-sectional area = .8816E+02 sq inch
```

```
TABULATION OF CALCULATED AVERAGE QUANTITIES:
Area weighted means for absolute values taken
over entire pipe (x,y) cross-sectional area
```

```
O   Axial flow velocity = .2565E+02 in/sec
O   Apparent viscosity = .5867E-04 lbf sec/sq in
O   Viscous stress, AppVis x dU/dx, = .6413E-03 psi
O   Viscous stress, AppVis x dU/dy, = .6308E-03 psi
O   Dissipation function = .2344E-01 lbf/(sec sq in)
O   Shear rate dU/dx = .1191E+02 1/sec
O   Shear rate dU/dy = .1162E+02 1/sec
O   Stokes product = .1604E-02 lbf/in
```

Second run:

```
POWER LAW FLOW OPTION SELECTED.
Power law fluid assumed, with exponent "n" equal
to .8000E+00 and consistency factor of .1000E-03
lbf sec^n/sq in.
```

```
Axial pressure gradient assumed as .3000E-01 psi/ft.
Total volume flow rate   = .2569E+04 gal/min
Cross-sectional area = .8816E+02 sq inch
```

```
TABULATION OF CALCULATED AVERAGE QUANTITIES:
Area weighted means for absolute values taken
over entire pipe (x,y) cross-sectional area
```

```
O   Axial flow velocity = .1013E+03 in/sec
O   Apparent viscosity = .4458E-04 lbf sec/sq in
O   Viscous stress, AppVis x dU/dx, = .1924E-02 psi
O   Viscous stress, AppVis x dU/dy, = .1892E-02 psi
O   Dissipation function = .2776E+00 lbf/(sec sq in)
O   Shear rate dU/dx = .4701E+02 1/sec
O   Shear rate dU/dy = .4587E+02 1/sec
O   Stokes product = .4813E-02 lbf/in
```

Third run:

```
POWER LAW FLOW OPTION SELECTED.
Power law fluid assumed, with exponent "n" equal
to .8000E+00 and consistency factor of .1000E-03
lbf sec^n/sq in.
```

```
Axial pressure gradient assumed as .5000E-01 psi/ft.
Total volume flow rate   = .4866E+04 gal/min
Cross-sectional area = .8816E+02 sq inch
```

```
TABULATION OF CALCULATED AVERAGE QUANTITIES:
Area weighted means for absolute values taken
over entire pipe (x,y) cross-sectional area
```

```
O   Axial flow velocity = .1918E+03 in/sec
O   Apparent viscosity = .3923E-04 lbf sec/sq in
O   Viscous stress, AppVis x dU/dx, = .3206E-02 psi
O   Viscous stress, AppVis x dU/dy, = .3154E-02 psi
O   Dissipation function = .8761E+00 lbf/(sec sq in)
O   Shear rate dU/dx = .8901E+02 1/sec
O   Shear rate dU/dy = .8686E+02 1/sec
O   Stokes product = .8022E-02 lbf/in
```

Fourth run:

```
POWER LAW FLOW OPTION SELECTED.
Power law fluid assumed, with exponent "n" equal
to .8000E+00 and consistency factor of .1000E-03
lbf sec^n/sq in.

Axial pressure gradient assumed as .1000E+00 psi/ft.
Total volume flow rate  = .1157E+05 gal/min
Cross-sectional area = .8816E+02 sq inch

TABULATION OF CALCULATED AVERAGE QUANTITIES:
Area weighted means for absolute values taken
over entire pipe (x,y) cross-sectional area

O  Axial flow velocity = .4561E+03 in/sec
O  Apparent viscosity = .3299E-04 lbf sec/sq in
O  Viscous stress, AppVis x dU/dx, = .6413E-02 psi
O  Viscous stress, AppVis x dU/dy, = .6308E-02 psi
O  Dissipation function = .4167E+01 lbf/(sec sq in)
O  Shear rate dU/dx = .2117E+03 1/sec
O  Shear rate dU/dy = .2066E+03 1/sec
O  Stokes product = .1604E-01 lbf/in
```

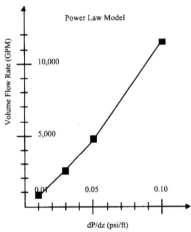

Figure 8-10. "Q vs. dp/dz" nonlinear behavior.

Simulation 5. Hydrate Plug, Herschel-Bulkley Flow in Circular Duct

---

In this set of runs, the "large blockage" example in Simulation 4 is reconsidered, with identical parameters, except that a nonzero yield stress of 0.005 psi is allowed. Thus, our "power law" fluid model becomes a "Herschel-Bulkley" fluid. Whereas smooth velocity distributions are typical of power law flows, e.g., Figure 8-9, the velocity field in flows with nonzero yield stress may contain "plugs" that move as solid bodies. For this simulation set, the plug flow velocity profiles obtained are typified by Figure 8-11.

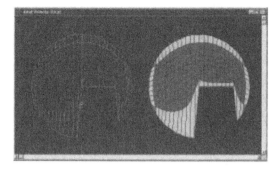

Figure 8-11. Plug flow in Herschel-Bulkley fluid.
(See CDROM for color slides.)

At 0.01 psi/ft, our flow rate is now obtained as 95.1 gpm, and at 0.10 psi/ft, we have 1,001 gpm. These flow rates are an order-of-magnitude below those calculated above; interestingly, the "Q vs. dp/dz" response in this example is almost linear, although this is not generally true for Herschel-Bulkley fluids. As before, we provide "typical numbers" in the tabulated results below, and also plot "Q vs. dp/dz" for what is an "exceptional" data set in Figure 8-12.

First run:

```
HERSCHEL-BULKLEY FLOW OPTION SELECTED.
Power law curve assumed with exponent "n" equal
to .8000E+00 and consistency factor "k" of .1000E-03
lbf sec^n/sq in.

Yield stress of .5000E-02 psi taken throughout.

Axial pressure gradient assumed as .1000E-01 psi/ft.
Total volume flow rate  = .9513E+02 gal/min
Cross-sectional area = .8816E+02 sq inch

Apparent viscosity and Stokes product set to
zero in plug regime for tabulation and display.

TABULATION OF CALCULATED AVERAGE QUANTITIES:
Area weighted means for absolute values taken
over entire pipe (x,y) cross-sectional area

O  Axial flow velocity = .3932E+01 in/sec
O  Viscous stress, AppVis x dU/dx, = .2042E-03 psi
O  Viscous stress, AppVis x dU/dy, = .1984E-03 psi
O  Dissipation function = .1446E-02 lbf/(sec sq in)
O  Shear rate dU/dx = .1180E+01 1/sec
O  Shear rate dU/dy = .1070E+01 1/sec
```

Second run:

```
HERSCHEL-BULKLEY FLOW OPTION SELECTED.
Power law curve assumed with exponent "n" equal
to .8000E+00 and consistency factor "k" of .1000E-03
lbf sec^n/sq in.

Yield stress of .5000E-02 psi taken throughout.

Axial pressure gradient assumed as .3000E-01 psi/ft.
Total volume flow rate = .2854E+03 gal/min
Cross-sectional area = .8816E+02 sq inch

Apparent viscosity and Stokes product set to
zero in plug regime for tabulation and display.

TABULATION OF CALCULATED AVERAGE QUANTITIES:
Area weighted means for absolute values taken
over entire pipe (x,y) cross-sectional area

O  Axial flow velocity = .1180E+02 in/sec
O  Viscous stress, AppVis x dU/dx, = .6126E-03 psi
O  Viscous stress, AppVis x dU/dy, = .5951E-03 psi
O  Dissipation function = .1302E-01 lbf/(sec sq in)
O  Shear rate dU/dx = .3539E+01 1/sec
O  Shear rate dU/dy = .3211E+01 1/sec
```

Third run:

```
HERSCHEL-BULKLEY FLOW OPTION SELECTED.
Power law curve assumed with exponent "n" equal
to .8000E+00 and consistency factor "k" of .1000E-03
lbf sec^n/sq in.

Yield stress of .5000E-02 psi taken throughout.

Axial pressure gradient assumed as .5000E-01 psi/ft.
Total volume flow rate = .4757E+03 gal/min
Cross-sectional area = .8816E+02 sq inch

Apparent viscosity and Stokes product set to
zero in plug regime for tabulation and display.

TABULATION OF CALCULATED AVERAGE QUANTITIES:
Area weighted means for absolute values taken
over entire pipe (x,y) cross-sectional area


O  Axial flow velocity = .1966E+02 in/sec
O  Viscous stress, AppVis x dU/dx, = .1021E-02 psi
O  Viscous stress, AppVis x dU/dy, = .9918E-03 psi
O  Dissipation function = .3616E-01 lbf/(sec sq in)
O  Shear rate dU/dx = .5899E+01 1/sec
O  Shear rate dU/dy = .5351E+01 1/sec
```

Fourth run:

```
HERSCHEL-BULKLEY FLOW OPTION SELECTED.
Power law curve assumed with exponent "n" equal
to .8000E+00 and consistency factor "k" of .1000E-03
lbf sec^n/sq in.

Yield stress of .5000E-02 psi taken throughout.

Axial pressure gradient assumed as .1000E+00 psi/ft.
Total volume flow rate  = .1001E+04 gal/min
Cross-sectional area = .8816E+02 sq inch

Apparent viscosity and Stokes product set to
zero in plug regime for tabulation and display.

TABULATION OF CALCULATED AVERAGE QUANTITIES:
Area weighted means for absolute values taken
over entire pipe (x,y) cross-sectional area

O  Axial flow velocity = .4085E+02 in/sec
O  Viscous stress, AppVis x dU/dx, = .2478E-02 psi
O  Viscous stress, AppVis x dU/dy, = .2463E-02 psi
O  Dissipation function = .2386E+00 lbf/(sec sq in)
O  Shear rate dU/dx = .1606E+02 1/sec
O  Shear rate dU/dy = .1637E+02 1/sec
```

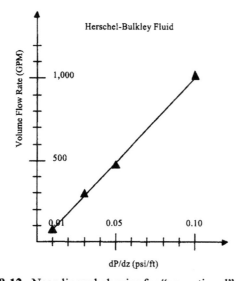

Figure 8-12. Near-linear behavior for "exceptional" data set.

Simulation 6. Eroding a Clogged Bed

Here, we start with the clogged pipe annulus of Chapter 7, where the inner pipe rests on the bottom, with sand almost filled to the top. We postulate a simple erosion model, where light particles are washed away at speeds greater than a given critical velocity. In the runs shown below, this value is always exceeded, so that the sand bed will always erode. In this final simulation set, the hole completely opens up, providing a successful conclusion to this chapter!

In order to provide general results, we again consider a Newtonian flow, so that the specific results given in the tabulations can be rescaled and recast more generally in the graph shown in Figure 8-14. While "Q vs. dp/dz" is linear in Newtonian fluids, note that "Q vs. N" is not. For that matter, even when a flow is Newtonian, the variation of Q versus any geometric parameter is typically nonlinear, and computational modeling will be required.

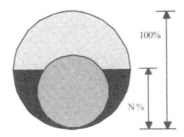

Figure 8-13. Clogged pipe simulation setup.

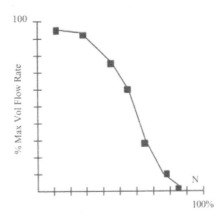

Figure 8-14. Generalized flow rate vs. dimensionless "fill-up."

In the following results, a unit cp Newtonian fluid is assumed, and a pressure gradient of 0.001 psi/ft is fixed throughout. A 6.4-inch diameter is taken for the outer circle, with "y = 0" referring to its center elevation; a "yheight" of -3.2 inches implies "no clogging," while + 2.0 is almost completely clogged. An inner 4.0-inch O.D. pipe rests at the very bottom of the annulus.

First run:

```
Enter YHEIGHT:  -3.2
Total volume flow rate  = .9340E+03 gal/min
Cross-sectional area = .2041E+02 sq inch

Area weighted means for absolute values taken
over entire pipe (x,y) cross-sectional area
O  Axial flow velocity = .1026E+03 in/sec
O  Apparent viscosity = .1465E-06 lbf sec/sq in
O  Viscous stress, AppVis x dU/dx, = .1951E-04 psi
O  Viscous stress, AppVis x dU/dy, = .2123E-04 psi
O  Dissipation function = .1226E-01 lbf/(sec sq in)
O  Shear rate dU/dx = .1331E+03 1/sec
O  Shear rate dU/dy = .1449E+03 1/sec
O  Stokes product = .1503E-04 lbf/in
```

Second run:

```
Enter YHEIGHT:  -2.2
Total volume flow rate  = .9383E+03 gal/min
Cross-sectional area = .1966E+02 sq inch

Area weighted means for absolute values taken
over entire pipe (x,y) cross-sectional area
O  Axial flow velocity = .1351E+03 in/sec
O  Apparent viscosity = .1465E-06 lbf sec/sq in
O  Viscous stress, AppVis x dU/dx, = .2453E-04 psi
O  Viscous stress, AppVis x dU/dy, = .2755E-04 psi
O  Dissipation function = .1635E-01 lbf/(sec sq in)
O  Shear rate dU/dx = .1674E+03 1/sec
O  Shear rate dU/dy = .1880E+03 1/sec
O  Stokes product = .1979E-04 lbf/in
```

Third run:

```
Enter YHEIGHT:  -1.2
Total volume flow rate  = .9157E+03 gal/min
Cross-sectional area = .1805E+02 sq inch

Area weighted means for absolute values taken
over entire pipe (x,y) cross-sectional area
O  Axial flow velocity = .1586E+03 in/sec
O  Apparent viscosity = .1465E-06 lbf sec/sq in
O  Viscous stress, AppVis x dU/dx, = .2506E-04 psi
O  Viscous stress, AppVis x dU/dy, = .3397E-04 psi
O  Dissipation function = .1947E-01 lbf/(sec sq in)
O  Shear rate dU/dx = .1710E+03 1/sec
O  Shear rate dU/dy = .2318E+03 1/sec
O  Stokes product = .2324E-04 lbf/in
```

Fourth run:

```
Enter YHEIGHT:  0.
Total volume flow rate  = .7837E+03 gal/min
Cross-sectional area = .1492E+02 sq inch

Area weighted means for absolute values taken
over entire pipe (x,y) cross-sectional area
O  Axial flow velocity = .1769E+03 in/sec
O  Apparent viscosity = .1465E-06 lbf sec/sq in
O  Viscous stress, AppVis x dU/dx, = .1963E-04 psi
O  Viscous stress, AppVis x dU/dy, = .4324E-04 psi
O  Dissipation function = .2234E-01 lbf/(sec sq in)
O  Shear rate dU/dx = .1340E+03 1/sec
O  Shear rate dU/dy = .2951E+03 1/sec
O  Stokes product = .2592E-04 lbf/in
```

Fifth run:

```
Enter YHEIGHT:  0.6
Total volume flow rate  = .6089E+03 gal/min
Cross-sectional area = .1253E+02 sq inch

Area weighted means for absolute values taken
over entire pipe (x,y) cross-sectional area
O  Axial flow velocity = .1714E+03 in/sec
O  Apparent viscosity = .1465E-06 lbf sec/sq in
O  Viscous stress, AppVis x dU/dx, = .1259E-04 psi
O  Viscous stress, AppVis x dU/dy, = .4737E-04 psi
O  Dissipation function = .2291E-01 lbf/(sec sq in)
O  Shear rate dU/dx = .8593E+02 1/sec
O  Shear rate dU/dy = .3233E+03 1/sec
O  Stokes product = .2511E-04 lbf/in
```

Sixth run:

```
Enter YHEIGHT:  1.2
Total volume flow rate  = .2823E+03 gal/min
Cross-sectional area = .8952E+01 sq inch

Area weighted means for absolute values taken
over entire pipe (x,y) cross-sectional area
O  Axial flow velocity = .1133E+03 in/sec
O  Apparent viscosity = .1465E-06 lbf sec/sq in
O  Viscous stress, AppVis x dU/dx, = .8603E-05 psi
O  Viscous stress, AppVis x dU/dy, = .3692E-04 psi
O  Dissipation function = .1379E-01 lbf/(sec sq in)
O  Shear rate dU/dx = .5871E+02 1/sec
O  Shear rate dU/dy = .2520E+03 1/sec
O  Stokes product = .1660E-04 lbf/in
```

Seventh run:

```
Enter YHEIGHT:  2.0
Total volume flow rate  = .5476E+02 gal/min
Cross-sectional area = .4458E+01 sq inch

Area weighted means for absolute values taken
over entire pipe (x,y) cross-sectional area
O  Axial flow velocity = .4484E+02 in/sec
O  Apparent viscosity = .1465E-06 lbf sec/sq in
O  Viscous stress, AppVis x dU/dx, = .4532E-05 psi
O  Viscous stress, AppVis x dU/dy, = .2281E-04 psi
O  Dissipation function = .5185E-02 lbf/(sec sq in)
O  Shear rate dU/dx = .3093E+02 1/sec
O  Shear rate dU/dy = .1557E+03 1/sec
O  Stokes product = .6570E-05 lbf/in
```

Eighth run:

```
Enter YHEIGHT:  2.5
Total volume flow rate  = .9648E+01 gal/min
Cross-sectional area = .2126E+01 sq inch

Area weighted means for absolute values taken
over entire pipe (x,y) cross-sectional area
O  Axial flow velocity = .1624E+02 in/sec
O  Apparent viscosity = .1465E-06 lbf sec/sq in
O  Viscous stress, AppVis x dU/dx, = .2321E-05 psi
O  Viscous stress, AppVis x dU/dy, = .1360E-04 psi
O  Dissipation function = .1832E-02 lbf/(sec sq in)
O  Shear rate dU/dx = .1584E+02 1/sec
O  Shear rate dU/dy = .9282E+02 1/sec
O  Stokes product = .2380E-05 lbf/in
```

Velocity field "snapshots" at different stages of the unclogging process are given in Figures 8-15a to 8-15f. Color slides of these figures are available on the enclosed CDROM. Although we have described the problem in terms of debris removal for eccentric annuli in horizontal drilling, it is clear that the computations are also relevant to wax removal in a simple bundled pipeline, where wax has formed at the top, when heat has been removed temporarily (the plots shown should then be turned upside-down).

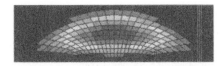

Figure 8-15a. Clogged annulus, "yheight" = 2.0 inches.

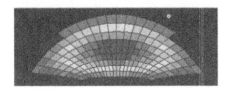

Figure 8-15b. Clogged annulus, "yheight" = 1.2 inches.

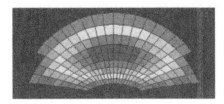

Figure 8-15c. Clogged annulus, "yheight" = 0.6 inches.

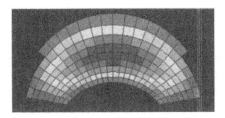

Figure 8-15d. Clogged annulus, "yheight" = 0.0 inches.

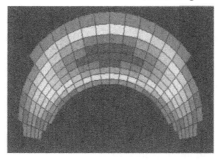

Figure 8-15e. Clogged annulus, "yheight" = -1.2 inches.

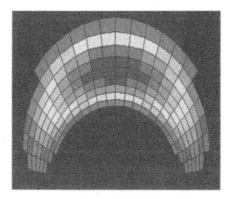

Figure 8-15f. *Unclogged* annulus, "yheight" = -2.2 inches.

The basic ideas on solids deposition and integrated non-Newtonian duct flow modeling have been developed in this chapter, and examples have been given that clearly demonstrate the dangers of even partial blockage.   In summary, minor blockage can significantly decrease flow rate, in a constant pressure gradient scenario. This also implies that minor blockages will require high start-up pressures when a pipeline system is recovering from stoppage. Here the problem can be severe, since temporary shutdowns generally allow blockages to solidify and adhere more securely. The "self-cleaning" ability of a flow is degraded, under the circumstances.

REFERENCES

Andersen, M.I, Isaksen, Ø., and Urdahl, O., "Ultrasonic Instrumentation for On-Line Monitoring of Solid Deposition in Pipes," SPE Paper No. 37437, Production Operations Symposium, Oklahoma City, March 1997.

Anderson, A.G., "Sedimentation," *Handbook of Fluid Mechanics*, Streeter, V.L., Editor, McGraw-Hill, New York, 1961.

Bern, P.A., Withers, V.R., and Cairns, R.J.R., "Wax Deposition in Crude Oil Pipelines," EUR Paper No. 206, European Offshore Petroleum Conference and Exhibition, London, England, October 1980.

Brown, T.B., Niesen, V.G., and Erickson, D.D., "Measurement and Prediction of the Kinetics of Paraffin Deposition," SPE Paper No. 26548, SPE Annual Technical Conference and Exhibition, Houston, October 1993.

Burger, E.D., Perkins, T.K., and Striegler, J.H., "Studies of Wax Deposition in the Trans Alaska Pipeline," *SPE Journal of Petroleum Technology*, June 1981.

Chang, C., Boger, D.V., and Nguyen, Q.D., "The Yielding of Waxy Crude Oils," *Industrial and Engineering Chemistry Research*, Vol. 37, No. 4, 1998, pp. 1551-1559.

Chang, C., Boger, D.V., and Nguyen, Q.D., "Influence of Thermal History on the Waxy Structure of Statically Cooled Waxy Crude Oils," SPE Paper No. 57959, to appear, *SPE Journal*.

Chang, C., Nguyen, Q.D., and Ronningsen, H.P., "Isothermal Start-Up of Pipeline Transporting Waxy Crude Oil," *Journal of Non-Newtonian Fluid Mechanics*, Vol. 87, 1999, pp. 127-154.

Chen, X.T., Butler, R.A., Volk, M., and Brill, J.P., "Techniques for Measuring Wax Thickness During Single and Multiphase Flow," SPE Paper No. 38773, SPE Annual Technical Conference and Exhibition, San Antonio, October 1997.

Cussler, E.L., *Diffusion Mass Transfer in Fluid Systems*, Cambridge University Press, 1997

Elphingstone, G.M., Greenhill, K.L., and Hsu, J.J.C., "Modeling of Multiphase Wax Deposition," *Journal of Energy Resources Technology*, Transactions of the ASME, Vol. 121, No. 2, June 1999, pp. 81-85.

Forsdyke, I.N., "Flow Assurance in Multiphase Environments," SPE Paper No. 37237, SPE International Symposium on Oilfield Chemistry, Houston, February 1996.

Hammami, A., and Raines, M., "Paraffin Deposition from Crude Oils: Comparison of Laboratory Results to Field Data," SPE Paper No. 38776, SPE Annual Technical Conference and Exhibition, San Antonio, October 1997.

Henriet, J.P., and Mienert, J., *Gas Hydrates: Relevance to World Margin Stability and Climatic Change*, The Geological Society, London, 1998.

Hsu, J.J.C., Santamaria, M.M., and Bribaker, J.P.: "Wax Deposition of Waxy Live Crudes under Turbulent Flow Conditions," SPE Paper No. 28480, SPE Annual Technical Conference and Exhibition, New Orleans, September 1994.

Hsu, J.J.C. and Brubaker, J.P., "Wax Deposition Measurement and Scale-Up Modeling for Waxy Live Crudes under Turbulent Flow Conditions," SPE Paper No. 29976, SPE International Meeting on Petroleum Engineering, Beijing, China, November 1995.

Hsu, J.J.C. and Brubaker, J.P., "Wax Deposition Scale-Up Modeling for Waxy Crude Production Lines," OTC Paper 7778, Offshore Technology Conference, Houston, May 1995.

Hunt, B.E., "Laboratory Study of Paraffin Deposition," *SPE Journal of Petroleum Technology*, November 1962, pp. 1259-1269.

Ibraheem, S.O., Adewumi, M.A., and Savidge, J.L., "Numerical Simulation of Hydrate Transport in Natural Gas Pipeline," *Journal of Energy Resources Technology*, Transactions of the ASME, Vol. 120, March 1998, pp. 20-26.

Jessen F.W. and Howell, J.N., "Effect of Flow Rate on Paraffin Accumulation in Plastic, Steel, and Coated Pipe," *Petroleum Transactions*, AIME, 1958, pp. 80-84.

Kapfer, W.H., "Flow of Sludges and Slurries," *Piping Handbook*, King, R., Editor, McGraw-Hill, New York, 1973.

Keating, J.F., and Wattenbarger, R.A., "The Simulation of Paraffin Deposition and Removal in Wellbores," SPE Paper No. 27871, SPE Western Regional Meeting, Long Beach, California, March 1994.

Majeed, A., Bringedal, B., and Overa, S., "Model Calculates Wax Deposition for North Sea Oils," *Oil and Gas Journal*, June 1990, pp. 63-69.

Matzain, A., "Single Phase Liquid Paraffin Deposition Modeling," MS Thesis, The University of Tulsa, Tulsa, Oklahoma, 1996.

Mendes, P.R.S. and Braga, S.L., "Obstruction of Pipelines during the Flow of Waxy Crude Oils," *Journal of Fluids Engineering*, Transactions of the ASME, December 1996, pp. 722-728.

Pate, C.B., "MMS/Deepstar Workshop on Produced Fluids," Offshore Technology Conference, Houston, May 1995.

Shauly, A., Wachs, A., and Nir, A., "Shear-Induced Particle Resuspension in Settling Polydisperse Concentrated Suspension," *International Journal of Multiphase Flow*, Vol. 26, 2000, pp. 1-15.

Singh P., Fogler H.S., and Nagarajan N., "Prediction of the Wax Content of the Incipient Wax-Oil Gel in a Pipeline: An Application of the Controlled-Stress Rheometer," *Journal of Rheology*, Vol. 43, No. 6, November-December 1999, pp. 1437-1459.

Souza Mendes, P.R., Braga, A.M.B., Azevedo, L.F.A., and Correa, K.S., "Resistive Force of Wax Deposits During Pigging Operations," *Journal of Energy Resources Technology*, Transactions of the ASME, Vol. 121, September 1999, pp. 167-171.

Wardhaugh, L.T., and Boger, D.V. "Flow Characteristics of Waxy Crude Oils: Application to Pipeline Design," *AIChE Journal*, Vol. 37, No. 6 June 1991, pp. 871-885.

Weingarten J.S., and Euchner, J.A., "Methods for Predicting Wax Precipitation and Deposition," *SPE Journal of Production Engineering*, February 1988, pp. 121-126.

9

Pipe Bends, Secondary Flows, and Fluid Heterogeneities

In this chapter, we develop the mathematical ideas needed to model flow in slow pipe bends, that is, in pipelines where the radius of curvature of the axis significantly exceeds duct cross-dimensions; in this limit, the corrections account for centrifugal effects only. For smaller radii of curvature, secondary viscous flows in the cross-plane appear, in addition to centrifugal effects; these are already discussed in experimental literature and are not reconsidered here.

Why study the large radius limit? In many drilling and subsea pipeline applications, slow geometric change along the axis is the rule, and secondary viscous flows are not anticipated. But when drillstrings turn horizontally, or when pipelines bend over mounds, flow differences at the top and bottom of the cross-section may initiate pipe blockage. Centrifugal effects may be associated with debris accumulations, not unlike observed preferential soil deposition occurring at selected sides of meandering river channels. How does solids deposition take place? Can local stress fields remove this debris?

These questions are eliciting interest from subsea pipeline engineers studying wax, hydrate, asphaltene, and sand removal. Thus, the calculation of velocity, shear rate, viscous stress and pressure drop for non-Newtonian flow in arbitrary ducts, in particular, clogged ducts, is important, and especially to steady-state and start-up operations. The work of this chapter, emphasizing the effects of bends, aims at supporting ongoing deposition studies. In the same way that different deposition mechanisms govern "cohesive" versus "non-cohesive particles" in sedimentary transport, we anticipate that different laws apply to different types of debris. Our methods use laboratory "n, k" data to predict field scale rheological properties, hopeful that these parameters will provide meaningful correlation variables for pipe clog prediction.

231

MODELING NON-NEWTONIAN DUCT FLOW IN PIPE BENDS

Bends in pipelines and annuli are interesting because they are associated with losses; that is, to maintain a prescribed volume flow rate, a greater pressure drop is required in pipes with bends than those without. This is true because the viscous stresses that act along pipe walls are higher. We will discuss the problem analytically, in the context of Newtonian flow; in this limit, exact solutions are derived for Poiseuille flow between curved concentric plates, but we will also focus on the form of the new differential equation used. The closed form expressions for Newtonian flow derived are new. Their derivation motivates our methodology for non-Newtonian flows in three-dimensional, curved, closed ducts, which can only be analyzed computationally.

Figure 9-1a. Viscous flow in a circular pipe.

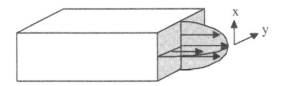

Figure 9-1b. Viscous flow in a rectangular duct.

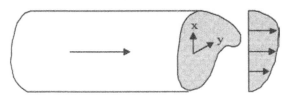

Figure 9-1c. Viscous flow in general duct.

Straight, closed ducts. We have so far derived analytical solutions for Newtonian flows in *straight* circular and rectangular conduits, as shown in Figures 9-1a and 9-1b. We also developed a general non-Newtonian viscous flow solver applicable to arbitrary *straight* ducts, e.g., see Figure 9-1c, utilizing curvilinear meshes, which supplements the algorithm of Chapter 2 applicable to eccentric annuli. For both ducts and annuli, we have demonstrated that the general solvers produce the correct velocity distributions in the simpler Newtonian and power law limits for ideal geometries. We now ask, "How are these methodologies extended to handle bends along duct and annular axes?" These extensions are motivated by the parallel plate solutions derived next.

Hagen-Poiseuille flow between planes. Let us consider here the plane Poiseuille flow between parallel plates shown in Figure 9-2a, but for simplicity, restrict ourselves first to Newtonian fluids.

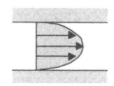

Figure 9-2a. Flow between parallel plates.

Let y be the coordinate perpendicular to the flow, with $y = 0$ and H representing the walls of a duct of height H. If $u(y)$ represents the velocity, the Navier-Stokes equations reduce to Equation 9-1, that is,

$$d^2u(y)/dy^2 = 1/\mu \; dP/dz \qquad (9-1)$$

$$u(0) = u(H) = 0 \qquad (9-2)$$

which is solved with the no-slip conditions in Equation 9-2. Again, μ is the Newtonian viscosity and dP/dz is the constant axial pressure gradient. The velocity solution is the well-known parabolic profile

$$u(y) = \tfrac{1}{2}(1/\mu \; dP/dz) \; y(y-H) \qquad (9-3)$$

which yields the volume flow rate "Q/L" (per unit length 'L' out of the page)

$$Q/L = \int_{0}^{H} u(y) \; dy = - (1/\mu \; dP/dz) \; H^3/12 \qquad (9-4)$$

Flow between concentric plates. Now suppose that the upper and lower walls are bent so that they conform to the circumferences of concentric circles with radii "R" and "R+H," where R is the radius of curvature of the smaller circle. We ask, "How are corrections to Equations 9-3 and 9-4 obtained?"

It is instructive to turn to Equation 3-18, the momentum law in the azimuthal "θ" direction introduced in Chapter 3 for rotating concentric flow. There, v_θ represented the velocity in the circumferential direction. We now draw upon that equation, but apply it to the flow between the concentric curved plates shown in Figure 9-2b.

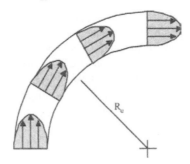

Figure 9-2b. Opened duct flow between concentric curved plates.

Since there is no flow perpendicular to the page, $v_z = 0$; also, $v_r = 0$ because the velocity is directed only tangentially, and F_θ is assumed to be zero. In these coordinates, the flow is steady, and θ and "z" variations vanish identically. Thus, Equation 3-18 reduces to the ordinary differential equation

$$d^2 v_\theta/dr^2 + 1/r \, dv_\theta/dr - v_\theta/r^2 = 1/\mu \, \{1/r \, dP/d\theta\} \qquad (9\text{-}5)$$

where the right side, containing the axial pressure gradient "$1/r \, dP/d\theta$," is approximately constant. It must be solved together with the no-slip conditions

$$v_\theta(R) = v_\theta(R+H) = 0 \qquad (9\text{-}6)$$

A closed form solution can be obtained as

$$\begin{aligned} v_\theta(r)/[1/\mu \, \{1/r \, dP/d\theta\}] &= \{(R+H)^3 - R^3\}/\{R^2 - (R+H)^2\} \times (r/3) \\ &\quad - R^2(R+H)^2/\{R^2 - (R+H)^2\} \times \{H/(3r)\} + 1/3 \, r^2 \qquad (9\text{-}7) \end{aligned}$$

Then, the volume flow rate "Q/L" per unit length (out of the page) is easily computed from

$$Q/L = \int_R^{R+H} v_\theta(r) \, dr \qquad (9\text{-}8)$$

exactly as

$$\begin{aligned} Q/L = & \ (1/18) \, (1/\mu \, dP/dz) \qquad (9\text{-}9) \\ & \{- 6R^3H - 9R^2H^2 - 5RH^3 - H^4 + 6R^4 \ln(R+H) \\ & + 12HR^3 \ln(R+H) + 6H^2R^2 \ln(R+H) - 6R^4 \ln(R) \\ & - 12R^3H \ln(R) - 6R^2H^2 \ln(R)\}/(2R+H) \end{aligned}$$

where we have replaced "$1/r \, dP/d\theta$" by "dP/dz." Now, in the limit $R \gg H$, Equation 9-9 simplifies to

$$Q/L \approx (1/\mu \, dP/dz) \, \{- H^3/12 + H^5/(180 \, R^2) + O(1/R^3)\} \qquad (9\text{-}10)$$

The first term in Equation 9-10 is the result in Equation 9-4, that is, the asymptotic contribution of the straight parallel plate solution. Subsequent terms represent corrections for finite R. In general, Equation 9-9 applies to all R and H combinations without restriction.

Typical calculations. It is interesting to ask, "How does total volume flow rate in such a curved 'pipe' compare with classical parallel plate theory?" For this purpose, consider the ratio obtained by dividing Equation 9-9 by Equation 9-4. It is plotted in Figure 9-2c, where we have set H = 1 and varied R.

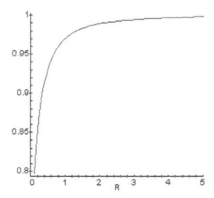

Figure 9-2c. Volume flow rate ratio, with H = 1.

This ratio tends to "1" quickly, when R > 5. We also ask, "What is the worst flow rate penalty possible?" If we take R → 0, it can be shown that the ratio approaches 2/3. Thus, for Newtonian flow between concentric plates, the volume flow rate is at worst equal to 2/3 of the value obtained between parallel plates for the same H. This assumes that the flow is steady and laminar, with no secondary viscous flow in the cross-sectional plane.

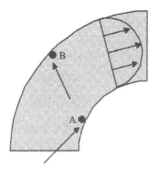

Figure 9-2d. Particles impinging at duct walls.

We also use the velocity solution in Equation 9-7 to study the viscous stresses at the walls of our concentric channel. Consider Figure 9-2d, which shows two impinging particles lodged at A and B, which may represent wax, hydrate, cuttings or other debris. The likelihood that they dislodge depends on the local viscous stress, among other factors. In this problem, $v_z = v_r = 0$ and $\partial/\partial\theta = \partial/\partial z = 0$, leaving the single stress component $\tau_{r\theta}(r) = \mu\, r\, \partial(v_\theta/r)/\partial r$. In particular, we plot "Stress Ratio" = $-\tau_{r\theta}(R+H)/\tau_{r\theta}(R)$ in Figure 9-2e, with H = 1 and varying R. The "minus" is used to keep the ratio positive, since the signs of the opposing stresses are opposite. The result is shown in Figure 9-2e.

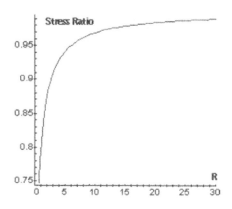

Figure 9-2e. "Stress Ratio," H = 1.

This graph shows that stresses at the outer wall are less whenever axis curvatures are finite. Thus, with all parameters equal, there is less likelihood that "B" will dislodge more quickly than "A." The velocity and stress solutions obtained here are also useful in determining how and where debris settle within the duct. Numerous factors enter, of course, among them, particle size, shape and distribution, buoyancy effects, local velocities and gradients, and so on. Such studies follow lines established in the sedimentary transport literature.

Flows in closed curved ducts. Our analysis showed that corrections for bends along the axis are obtained by solving $v_\theta(r)$ in cylindrical coordinates. It is apparent that the extension of Equations 9-9 and 9-10 to cover *closed* rectangular ducts (versus "opened" concentric plates) with finite radius of curvature, e.g., Figure 9-1b, requires the solution of Equation 9-5 *with* the "$\partial^2 v_\theta/\partial z^2$" term in Equation 3-18, leading to the partial differential equation

$$\partial^2 v_\theta/\partial r^2 + 1/r\; \partial v_\theta/\partial r - v_\theta/r^2 + \partial^2 v_\theta/\partial z^2 = 1/\mu\; \partial P/\partial z \qquad (9\text{-}11)$$

For the duct in Figure 9-1b, the no-slip conditions are $v_\theta(R,z) = v_\theta(R+H,z) = 0$ and $v_\theta(r,z_1) = v_\theta(r,z_2) = 0$, where $z = z_1$ and z_2 are end planes parallel to the page.

With our extension to rectangular geometries clear, the passage to *bent ducts with arbitrary closed cross-sections*, e.g., Figure 9-3, is obtained by taking Equation 9-11 again, but with no-slip conditions applied along the perimeter of the shaded area. Ducts with multiple bends are studied by combining multiple ducts with piecewise constant radii of curvature. Since the total flow rate is fixed, each section will be characterized by different axial pressure gradients.

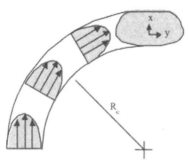

Figure 9-3. Arbitrary closed duct with curved axis.

Of course, Equation 9-11 is quite different from Equation 2-31, that is, from "$\partial^2 u/\partial y^2 + \partial^2 u/\partial x^2 \approx N(\Gamma)^{-1}\ \partial P/\partial z$" solved for straight ducts and annuli. In order to use the previous algorithm, we rewrite Equation 9-11 in the form

$$\partial^2 v_\theta/\partial r^2 + \partial^2 v_\theta/\partial z^2 = 1/\mu\ \partial P/\partial z - 1/R\ \partial v_\theta/\partial r + v_\theta/R^2 \qquad (9\text{-}12)$$

where we have transferred the new terms to the right side, and replaced the variable "r" coefficients by constants, assuming $R \gg H$ so that $r \approx R$.

The "r, z" in Equation 9-12 are just the "y, x" cross-sectional variables used earlier. In our iterative solution, the right side velocity terms of Equation 9-12 are evaluated using latest values, with the relaxation method continuing until convergence. For non-Newtonian flows, "μ" is replaced by $N(\Gamma)$ to a first approximation. These changes are easily implemented in software. For example, in Chapters 2 and 7, our "line relaxation" Fortran source code included the lines,

```
     WW(J)  =  -ALPHA(I,J)*(U(I-1,J)+U(I+1,J))/DPSI2
  1            +GAKOB(I,J)*GAKOB(I,J)* PGRAD/APPVIS(I,J)
  2            +2.0*BETA(I,J)*
  3            (U(I+1,J+1)-U(I-1,J+1)-U(I+1,J-1)+U(I-1,J-1))/
  4            (4.*DPSI*DETA)
```

which incorporate "$N(\Gamma)^{-1}\ \partial P/\partial z$," where the other terms shown are related to the Thompson mapping. To introduce pipe curvature, the bolded term is simply replaced as follows,

```
      CHANGE = PGRAD/APPVIS(I,J)
    1        -(YETA(I,J)*(U(I+1,J)-U(I-1,J))/(2.*DPSI)
    2        - YPSI(I,J)*(U(I,J+1)-U(I,J-1))/(2.*DETA))/
    3        (GAKOB(I,J)*RCURV)
    4        + U(I,J)/(RCURV**2.)
C
      WW(J) = -ALPHA(I,J)*(U(I-1,J)+U(I+1,J))/DPSI2
    1        +GAKOB(I,J)*GAKOB(I,J)*CHANGE
    2        +2.0*BETA(I,J)*
    3        (U(I+1,J+1)-U(I-1,J+1)-U(I+1,J-1)+U(I-1,J-1))/
    4        (4.*DPSI*DETA)
```

The second and third lines of our Fortran source code for "CHANGE" represent the "r" velocity derivative in transformed coordinates.

Newtonian calculations similar to those performed for "concentric plate Pouseuille flow" show that, when pressure gradient is prescribed, volume flow rate again decreases as the radius of curvature R_c tends to zero. For a circular cross-section of radius R, the decrease is roughly 20% relative to Hagen-Poiseuille flow when R_c and R are comparable. We have focused on Newtonian flows because exact solutions were available, and importantly, our results applied to all viscosities and pressure gradients. However, results will vary for pipelines with non-circular cross-sections, non-Newtonian flow, or both; general conclusions cannot be offered, of course, but computations can now be easily performed with the numerically stable implementation derived above.

FLUID HETEROGENEITIES AND SECONDARY FLOWS

The term "secondary flow" is often used in different contexts. In the first, it refers to small scale physical features embedded within simpler larger flows; in another, it refers to effects that cannot be modeled because important terms have been neglected in the formulation. In this section, cautionary advice is offered to readers: unlike finite element simulators for structural analysis, general purpose computational fluid-dynamic solvers rarely support the broad range of physics encountered in reality, and their use is not encouraged.

For example, consider the flow past the rearward step shown in Figure 9-4. The left schematic illustrates attached streamline solutions for "ideal flows" satisfying Laplace's equation, while the right figure shows separation eddies obtained by solving the full Navier-Stokes equations, in this case, using the vorticity-streamfunction method. The flow past a circle, assuming ideal flow, would show symmetric fore and aft streamline patterns. But more realistically, the solution in Figure 9-5 is obtained. At higher Reynolds numbers, unsteady patterns with downstream "shed" vortices are observed, which cannot be modeled unless time dependency is further allowed in the governing equations.

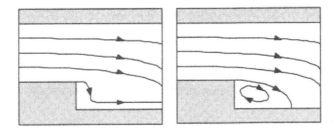

Figure 9-4. Flow past rearward step, "ideal" at left, actual at right.

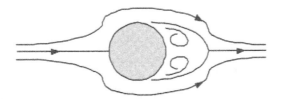

Figure 9-5. Viscous flow past a circle.

Our point is clear: in spite of the mathematical sophistication offered here, and by others in similar investigations, most computed fluid-dynamical solutions describe but a narrow aspect of the general physical problem. For annular and pipeline flows, we have seen in Chapter 4 that density gradients can lead to recirculating vortex zones, for example, as in Figure 9-6 below. These effects, which arise from density segregation, are not included in our pipe bend models.

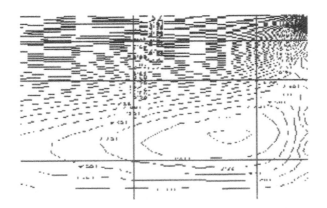

Figure 9-6. Recirculation zones due to density segregation.

And neither does our analysis include the secondary viscous flows observed in the cross-planes of pipes with small radii of curvature, as in Figure 9-7. These exist because fluid particles near the flow axis, which have higher velocities, are acted upon by larger centrifugal forces than slower particles near the walls. This gives rise to a secondary flow pattern that is directed outwards at the center and inwards near the wall. The reader should consider which flow(s) pertain to his problem before using formulas or commercial software models.

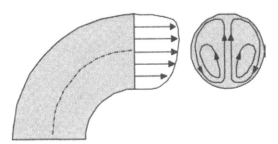

Figure 9-7. Counter-rotating viscous flow.

The characteristic dimensionless variable in laminar Newtonian flow is the "Dean number" $D = \frac{1}{2}$ (Reynolds number) $\sqrt{}$ (Pipe radius/Radius of curvature) for the secondary flow shown. This flow is associated with losses in addition to the above centrifugal ones because additional energy is imparted to rotation. References to fundamental studies appear in Schlichting (1968), and recent studies including swirl and turbulence are cited below. Roache (1972) gives an excellent "hands on" summary of basic simulation algorithms for rectangular grids, although the work addresses Newtonian flows. These methods were extended to curvilinear meshes and nonlinear rheologies in Chapters 2 and 7.

REFERENCES

Anwer, M., "Rotating Turbulent Flow Through a 180 Degree Bend," Ph.D. Thesis, Arizona State University, 1989.

Moene, A.F., Voskamp, J.H., and Nieuwstadt, F.T.M., "Swirling Pipe Flow Subject to Axial Strain," Advances in Turbulence VII. Proceedings of the Seventh European Turbulence Conference, 1998, pp. 195-198.

Parchen, R.R., and Steenbergen, W., "An Experimental and Numerical Study of Turbulent Swirling Pipe Flows," *J. Fluid Eng.* 120, 1998, pp. 54-61.

Roache, P.J., *Computational Fluid Dynamics*, Hermosa Publishers, Albuquerque, New Mexico, 1972.

Schlichting, H., *Boundary Layer Theory*, McGraw-Hill, New York, 1968.

Steenbergen, W., and Voskamp, J.H., "The Rate of Decay of Swirl in Turbulent Pipe Flow," *Flow Measurement. and Instrumentation*, 9, 1998, pp. 67-78.

10

Advanced Modeling Methods

In this final chapter, we describe some research areas that will extend the power of our numerical models, making them even more practical for routine use. These subject areas include the treatment of "complicated" flow domains, convergence acceleration, fast solutions to Laplace's equation, and the use of rheological models special to petroleum applications. Finally, we comment on the curvilinear grid simulation models that are presently available, and various software features that have been developed to make the programs easy to use.

COMPLICATED PROBLEM DOMAINS

In this section, we will provide a unified picture of boundary value problem formulation on complicated domains. Although this is discussed in the context of non-Newtonian fluid flow in pipelines and annuli, we emphasize that the ideas apply generally to all problems in continuum mechanics. For example, we have studied axial flows of power law fluids satisfying

$$N(\Gamma) = k \ [\ (\partial u/\partial y)^2 + (\partial u/\partial x)^2 \ ]^{(n-1)/2} \qquad (2\text{-}29)$$

$$\partial^2 u/\partial y^2 + \partial^2 u/\partial x^2 \approx N(\Gamma)^{-1} \ \partial P/\partial z \qquad (2\text{-}31)$$

subject to no-slip "$u = 0$" velocity boundary conditions at solid walls. But we could just as easily solve "$\partial^2 T/\partial y^2 + \partial^2 T/\partial x^2 = 0$" for steady-state heat transfer, specifying nonzero T's at prescribed boundaries, as we have in Chapter 6. Extension to transient problems is straightforward. For instance, the unsteady equation "$\partial^2 T/\partial y^2 + \partial^2 T/\partial x^2 = \alpha \ \partial T/\partial t$" can be solved by explicit or implicit finite difference time integration in the physical plane (y,x) using standard computational techniques. If we map or transform this equation into (r,s)

241

coordinates, the revised model takes an identical form, except that α is replaced by α *times* a stored function related to the Jacobian of the transformation. Then, existing integration methods can be used; however, boundary conditions can now be satisfied exactly on complicated domains. To keep our ideas focused on the subject of this book, let us return to pipeline and annular flow applications.

The author first studied annular flow in *Borehole Flow Modeling*, and extended the methods to duct flow in Chapters 7 and 8 of this book. This chronological order may be confusing to new readers exposed to the subject for the first time. In order to explain how our methods can be generalized to complicated flow domains, it is convenient to discuss duct flows first, annular flows next, and finally, general pipeline bundles with complicated annuli formed by two or more internal "holes" or pipes.

Singly-connected regions. In this book, ducts need not be circular or rectangular cross-sectionally; in fact, they may take any of the forms shown in Figure 10-1a. That is, they may have "no sides," "three sides," or any number of sides: the number of "sides" is irrelevant topologically. But how are these domains mapped into the computational rectangle in Figure 10-1b? What is the exact "recipe" used to generate velocity flowfields? Let us consider the physical plane first, and perform the following sequence of steps. (1) Select "corner" or "vertex" points A, B, C, and D. (2) Choose like numbers of spaced nodes along AB and DC, and AD and BC, and record their (y,x) positions. (3) These nodes, importantly, need not be equally spaced, and their numbers in the "vertical" and "horizontal" directions need not be equal.

Now let us turn our attention to Figure 10-1b, and take the following actions. (1) Assign the previous (y,x) values along the corresponding line segments of the rectangle, taking care to maintain the same order or clockwise sense. For convenience, these values can be assigned at equally spaced intervals in the (r,s) plane. (2) The values in the previous step provide the boundary conditions for the transformations in Equations 2-36 and 2-37, that is,

$$(y_r^2 + x_r^2)\, y_{ss} - 2(y_s y_r + x_s x_r)\, y_{sr} + (y_s^2 + x_s^2)\, y_{rr} = 0 \qquad (2\text{-}36)$$

$$(y_r^2 + x_r^2)\, x_{ss} - 2(y_s y_r + x_s x_r)\, x_{sr} + (y_s^2 + x_s^2)\, x_{rr} = 0 \qquad (2\text{-}37)$$

as before. (3) Discretize the foregoing equations using central difference approximations and solve the resulting system using finite difference relaxation methods, for example, as outlined in Chapter 7. (4) Once the functions $x(r,s)$ and $y(r,s)$ are known, they are tabulated, and used to evaluate the coefficients in the transformed flow equation, namely,

$$(y_r^2 + x_r^2)\, u_{ss} - 2(y_s y_r + x_s x_r)\, u_{sr}$$
$$+ (y_s^2 + x_s^2)\, u_{rr} \approx (y_s x_r - y_r x_s)^2\, \partial P / \partial z\, / N(\Gamma) \qquad (2\text{-}38)$$

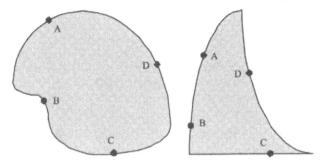

Figure 10-1a. "Singly-connected," duct flow domains.

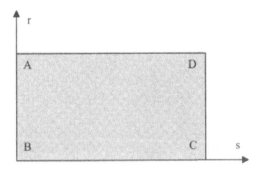

Figure 10-1b. Simple rectangle without "branch cuts."

Next, (5) the axial velocity u(r,s) in the transformed plane is obtained by solving Equation 2-38 using a similar finite difference method. (6) Then, the velocity u(y,x) in the physical plane must be displayed using suitable graphical techniques. (7) Finally, shear rates are computed from Equations 2-39 and 2-40, from which apparent viscosities and viscous shear stresses are obtained as outlined in Chapter 2.

$$u_y = (x_r u_s - x_s u_r)/(y_s x_r - y_r x_s) \qquad (2\text{-}39)$$

$$u_x = (y_s u_r - y_r u_s)/(y_s x_r - y_r x_s) \qquad (2\text{-}40)$$

Doubly-connected regions. A flow annulus, no more than a "duct with a hole," can be similarly considered. In Figure 10-2a, we have "cut" our annulus by introducing "branch cuts" B_1 and B_2. These fictitious lines need not be straight, but they are infinitesimally close. The result is a "singly-connected" duct domain, as opposed a "doubly-connected" annulus, and can be modeled as such provided we define suitable boundary conditions along B_1 and B_2. Now, the physical boundary conditions in Figure 10-2a, e.g., "no-slip" velocity, along C_1 and C_2, apply to the same line segments in Figure 10-2b.

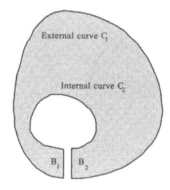

Figure 10-2a. "Doubly-connected," annular domain with "branch cuts."

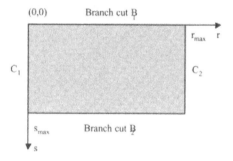

Figure 10-2b. Simple mapped rectangular space.

But how do we handle the two new branch cuts? Since B_1 and B_2 share the same discrete sets of (y,x) values chosen by the user to represent them, these identical (y,x) values are used as boundary conditions for both upper and lower horizontal lines. Then, Equations 2-36 and 2-37 are solved as before. The solution of Equation 2-38 for velocity is different. The transformed problem, formulated in the plane of Figure 10-2b, applies no-slip conditions along C_1 and C_2. How are B_1 and B_2 treated now? Note, each point within the physical annulus must possess a single unique speed. Thus, corresponding (y,x) points along B_1 and B_2 have identical values. Since these "u" values are not known until the iterations converge, those used in the program are the values obtained from the previous iteration; on convergence, "u" will vary along a branch cut consistently with the flow equations. This "single-valuedness" is a physical requirement Not all problems have single-valued solutions. For example, in heat transfer applications, branch cuts can be placed along insulators, across which two distinct temperatures may coexist. A similar analogy is found in Darcy flow past thin shale obstacles, which support different pressures at opposing sides. These are mathematically known as "double-valued" solutions.

Triply-connected regions. Now consider axial flow in an annulus with two "holes," as in Figure 10-3a for a pipe bundle concept slightly more complicated than that of Chapter 6. To keep the schematic simple, two smaller circles are shown centered within a larger one; we emphasize that there is no requirement that any of these closed curves be circular or aligned. As in the previous example, our approach is the use of "branch cuts" that transform the problem into one for simple duct flow, as Figure 10-3b clearly shows.

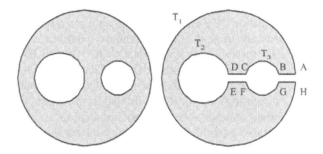

Figure 10-3a,b. "Triply-connected" region, before and after branch cuts.

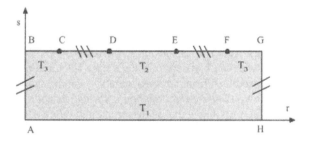

Figure 10-3c. Mapped domain for triply-connected domains.

In Figure 10-3b, we have indicated boundary conditions T_1, T_2 and T_3 as shown, for example, describing a bundled pipeline heat transfer problem with dual heating cylinders. These temperature boundary conditions map into line segments AH, DE, BC, and FG as shown in Figure 10-3c. Like "tick marks" indicate the "single-valuedness" conditions applied here. For example, solutions along CD are identical to those on EF, and similarly, for AB and HG. Similar considerations apply to "n-connected" regions. These topological ideas, incidentally, are extremely relevant to Darcy flow in petroleum reservoirs, which typically contain numerous "holes," otherwise known as oil wells. In this context, we emphasize that boundary conforming, curvilinear meshes permit detailed flow resolution, more than enough to resolve the local logarithmic pressure singularities that cannot be captured on cruder Cartesian meshes.

CONVERGENCE ACCELERATION

Here we discuss fast methods for irregular grid generation. Existing solution methods solving $x(r,s)$ and $y(r,s)$ "stagger" the solutions for Equations 2-36 and 2-37. Crude solutions are used to initialize the coefficients of Equation 2-36, and improvements to $x(r,s)$ are obtained. These are used to evaluate the coefficients of Equation 2-37, in order to obtain an improved $y(r,s)$; then, attention turns to Equation 2-36 again, and so on, until convergence is achieved.

Various means are used to implement these iterations, as noted in the review paper of Thompson (1984), e.g., "point SOR," "line SLOR," "line SOR with explicit damping," "alternating-direction-implicit," and "multigrid," with varying degrees of success. Often these schemes diverge computationally. In any event, the "staggering" noted above introduces different "artificial time levels" while iterating, however, classic numerical analysis suggests that faster convergence and improved stability is possible by reducing their number.

A new approach to solve Thompson's equations rapidly was proposed by Chin (2000) and based on a very simple idea. Consider "$z_{rr} + z_{ss} = 0$," for which "$z_{i,j} \approx (z_{i-1,j} + z_{i+1,j} + z_{i,j-1} + z_{i,j+1})/4$" holds on constant grids. This averaging law was derived earlier for flow in rectangular ducts. It motivates the *recursion formula* $z_{i,j}{}^n = (z_{i-1,j}{}^{n-1} + z_{i+1,j}{}^{n-1} + z_{i,j-1}{}^{n-1} + z_{i,j+1}{}^{n-1})/4$ often used to illustrate and develop "multi-level" iterative solutions; an approximate, and even trivial solution, can be used to initialize the calculations, and nonzero solutions are always produced from nonzero boundary conditions.

But the well known *Gauss-Seidel* method is fastest: as soon as a new value of $z_{i,j}$ is calculated, its previous value is discarded and overwritten by the new value. This speed is accompanied by low memory requirements, since there is no need to store both "n" and "n-1" level solutions: only a single array, $z_{i,j}$ itself, is required in programming. Our new approach to Equations 2-36 and 37 was motivated by this simple idea. Rather than solving for $x(r,s)$ and $y(r,s)$ in a "staggered, leap-frog" manner, is it possible to update x and y *simultaneously* in a similar "once only" manner? Is convergence significantly increased? How do we solve in Gauss-Seidel fashion? What are the programming implications?

Complex variables provides the vehicle. We *define* a dependent variable "z" by $z(r,s) = x(r,s) + i\, y(r,s)$, in particular, adding Equation 2-36 plus i times Equation 2-37, to obtain $(x_r^2 + y_r^2)\, z_{ss} - 2(x_s x_r + y_s y_r)\, z_{sr} + (x_s^2 + y_s^2)\, z_{rr} = 0$. Now, the "complex conjugate" of "z" is $z^*(r,s) = x(r,s) - i\, y(r,s)$, from which we find $x = (z + z^*)/2$ and $y = -i\,(z - z^*)/2$. Substitution produces the simple and equivalent "one equation" result

$$(z_r\, z_r^*)\, z_{ss} - (z_s\, z_r^* + z_s^*\, z_r)\, z_{sr} + (z_s\, z_s^*)\, z_{rr} = 0 \qquad (10\text{-}1)$$

This form yields significant advantages. First, when "z" is declared as a complex variable in a Fortran program, Equation 10-1 represents, for all practical purposes, a *single* equation in z(r,s). There is now no need to "leap frog" between x and y solutions at all, since a single formula that is analogous to "$z_{i,j} = (z_{i-1,j} + z_{i+1,j} + z_{i,j-1} + z_{i,j+1})/4$" is easily written for $z_{i-1,j}$ using second-order accurate central differences. Because both x and y are simultaneously resident in computer memory, the "extra" time level present in staggered schemes is completely eliminated, as in the Gauss-Seidel method.

In hundreds of test simulations conducted using point and line relaxation, convergence times are shorter by factors of two to three, with convergence rates far exceeding those obtained for cyclic solutions between x(r,s) and y(r,s). Convergence appears to be unconditional, monotonic, and stable. Because Equation 10-1 is nonlinear, von Neumann tests for exponential stability and traditional estimates for convergence rate do not apply, but our evidence for stability and convergence, while purely empirical so far, remains strong and convincing. In fact, we have not encountered a single divergent simulation.

FAST SOLUTIONS TO LAPLACE'S EQUATION

In Chapter 6, we discussed the use of "heat pipes" in raising annular flow temperature, which reduces the probability of wax and hydrate formation. In the simplest limit, Laplace's equation for the temperature T arises, when the effects of heat conduction are dominant. In practice, an operator may want to increase source temperatures in response to increases in crude viscosity. Does this mean that the temperature problem must be solved anew? The answer is "No." In fact, once our mappings are obtained and stored, there is *never* a need to solve Laplace's equation subject to several types of boundary conditions*!*

The grid generation methods used so far are powerful, and made even more versatile with convergence acceleration. *However, the best is yet to come*: solutions to several classes of steady-state boundary value problems are "automatic" and "free" in a literal sense*!* Very little additional work is required to obtain their solution. Thus, numerous practical problems can be solved in the field, with very little computational power required.

In the aerospace industry, the x(r,s) and y(r,s) define coordinates that might host the Navier-Stokes or the simpler boundary layer equations. In many oilfield applications, Laplace's equation arises on doubly-connected domains, and different values of the dependent variable are applied at inner and outer contours. For instance, in petroleum engineering, steady liquid Darcy flows satisfy "$p_{xx} + p_{yy} = 0$," while in annular pipe bundles, temperature fields are determined from "$T_{xx} + T_{yy} = 0$," as shown in Figure 10-4.

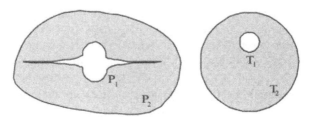

Figure 10-4. Practical formulations in (x,y) physical plane.

The conventional wisdom is easily summarized. Grid generation and temperature (or pressure) analysis are "obviously" independent and sequential tasks: *first* create the grid system, *then* obtain pressure. However, we can show that *multiple temperature solutions* are available without further effort once the grid is generated! Under conformal transformation, "$T_{xx} + T_{yy} = 0$" becomes "$T_{rr} + T_{ss} = 0$" for T(r,s). But we do *not* need to solve numerically for T(r,s), because an analytical solution is easily obtained.

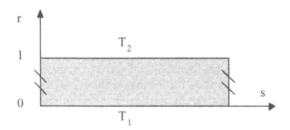

Figure 10-5. Simplified formulation in (r,s) plane.

To see why, observe from Figure 10-5 that, because the boundary values T_1 and T_2 are constant, the solution T(r,s) must be independent of s. Thus, our "$T_{rr} + T_{ss} = 0$" becomes the ordinary differential equation $T_{rr} = 0$, whose solution is just a linear function of "r." In summary, if inner and outer boundary values $T_1(t)$ and $T_2(t)$ are prescribed at r = 0 and 1, where "t" denotes a possible parametric dependence on time, the required solution is

$$T(r,s;t) = (T_2 - T_1) \, r(x,y) + T_1 \tag{10-2}$$

which is a linear function of r(x,y) alone! In other words, once x(r,s) and y(r,s) are available, and inverted (by table) to give s = s(x,y) and r = r(x,y), the complete solution to the "temperature-temperature" boundary value problem is available by *rescaling* using Equation 10-2. Thus, there is *never* a need to "solve for temperature" since our grid generation problem already contains the ingredients needed to construct the solution. This makes our techniques particularly useful for real-time control software. Boundary value problems can be "solved" with little effort beyond simply storing and rescaling arrays.

We have demonstrated the availability of "free" solutions for simple "temperature-temperature" problems, but this availability applies also to "temperature-heat flux" formulations. To show this, let us introduce a normalized function $T^* = 1 + (T - T_2)/(T_2 - T_1)$ that similarly satisfies "$T_{rr} = 0$." This function is "1" and "0" along the outer and inner reservoir contours, and is independent of T_2 and T_1. Now, the flux F through any closed curve surrounding the inside contour is

$$F = \alpha \int \nabla T \cdot \mathbf{n} \, dl \qquad (10\text{-}3)$$

where $\mathbf{n}$ is the outward unit normal, dl is an incremental length, and α is a constant. Since T_1 can be taken across the integral, simple algebra shows that

$$F = \alpha (T_2 - T_1)/I \qquad (10\text{-}4)$$

where the integral

$$I = \int \nabla(T)^* \cdot \mathbf{n} \, dl = \int \nabla r \, (x,y) \cdot \mathbf{n} \, dl \qquad (10\text{-}5)$$

depends only on geometrical details. It is calculated once and for all, and is presumed to be known. Hence, our algorithm: once F and either of T_2 and T_1 are specified, the remaining temperature is obtained by solving Equation 10-4, which is simple algebra. With both T_2 and T_1 now available, the corresponding temperature distribution is obtained directly from Equation 10-2.

In summary, the "recipe" for solving general flux problems is easily stated. First, solve the mapping problem for r(x,y) and s(x,y), then store r(x,y) and calculate "I." Next, specify any two of F, T_2 and T_1 , and solve Equation 10-4 for the remaining parameter. Finally, calculate the spatial temperature distribution from Equation 10-2. We emphasize that inner and outer contours may take any shape, and since our approach applies to any problem satisfying Laplace's equation, the applications of the new method are broad. In a sense, the "r(x,y)" used here plays the role of the "x," "log r," and "1/r" elementary solutions used in linear, cylindrical and spherical Laplace equation methods.

SPECIAL RHEOLOGICAL MODELS

The rheology literature focuses on several well-known models, e.g., power law, Bingham plastic, Herschel-Bulkley, and Casson, but very often, these models are not adequate for petroleum applications. The oil industry continually improves its rheological models through detailed laboratory work, and in this section, we will describe some of the industry's specialized needs.

Foam flows. One area of strong interest is "foam flow" in "underbalanced drilling." In underbalanced drilling, the hydrostatic pressure of the drilling fluid is maintained below the formation pore pressure. This permits higher rates of penetration, minimized formation damage, and reduced flow losses. Our knowledge of foam rheology is presently incomplete. Essentially, a "foam" is a highly compressible dispersion of gas bubbles in a continuous liquid matrix, strongly affected by temperature and pressure. Its apparent viscosity is a function of its "quality" (ratio of gas to total volume), bubble size distribution, and polymer presence. There is no clear consensus regarding the applicability of the traditional rheology models, e.g., the existence of yield stress has been suggested. However, there is overall agreement that macroscopic "velocity slip" at solid boundaries represents a new modeling parameter. In our research efforts, the no-slip condition is replaced by functional relationships of the form "u = f($\partial u/\partial x$, $\partial u/\partial y$, P, T)," using guidelines offered in (Karynik, 1988). The following references are especially useful:

- Edwards, D.A., Brenner, H., and Wasan, D.T., *Interfacial Transport Processes and Rheology*, "Chapter 14, Foam Rheology," Butterworth-Heinemann Series in Chemical Engineering, Boston, 1991.

- Edwards, D.A., and Wasan, D.T., "Foam Dilatational Rheology, I: Dilatational Viscosity," *J. Colloid Interface Science*, Vol. 139, 1990, pp. 479-487.

- Khan, S.A., and Armstrong, R.C., "Rheology of Foams, I: Theory for Dry Foams," *J. Non-Newtonian Fluid Mechanics*, Vol. 22, 1987, pp. 1-22.

- Kraynik, A.M., "Foam Flows," in *Annual Review of Fluid Mechanics*, Vol. 20, Annual Reviews, Inc., Palo Alto, 1988, pp. 325-357.

- Saintpere, S., Herzhaft, B., and Toure, A., "Rheological Properties of Aqueous Foams for Underbalanced Drilling," SPE Paper No. 56633, SPE Annual Technical Conference and Exhibition, Houston, October 1999.

Drilling applications. Borehole cleaning motivated our early rheology modeling, in which we made extensive use of "n" and "k" values reported by the University of Tulsa. These properties characterized muds used in the late 1980s. With recent interest in deep subsea applications increasing, the effect of low temperature and high pressure on drilling fluid properties must be determined. The following articles provide an introduction to the subject.

- Davison, J.M., Clary, S., Saasen, A., Allouche, M., Bodin, D., Nguyen, V.A., "Rheology of Various Drilling Fluid Systems Under Deepwater Drilling Conditions and the Importance of Accurate Predictions of Downhole Fluid Hydraulics," SPE Paper No. 56632, SPE Annual Technical Conference and Exhibition, Houston, 1999.

- Rommetveit, R., and Bjorkevoll, K.S., "Temperature and Pressure Effects on Drilling Fluid Rheology and ECD in Very Deep Wells," SPE/IADC Paper No. 39282, 1997 SPE/IADC Middle East Drilling Technology Conference, Bahrain, November 1997.

Other oilfield applications. Of particular significance to stimulation and hydraulic fracturing is "proppant transport rheology." Proppants, of course, carry solid particles that "prop" open fractures, thereby enhancing reservoir production by increasing exposed surface area. It is important to understand the behavior of fracturing fluids under realistic downhole conditions, and to correlate this with observed carrying capacities as a function of shear history. Research should identify rheological properties that improve proppant placement. Efforts to predict fracture fluid properties, for example, as functions of gel, crosslinker, and breaker type, concentration versus shear, heat-up profile, and time at temperature, will make fracturing less empirical and more predictive. The measurement of density stratification represents another area of interest. Recent observations in different areas of drilling and process engineering have identified problematic recirculation vortexes of the kind studied in Chapter 4. These also represent potential problem areas in deep sea applications, where gravity segregation remains a definite possibility.

Subsea pipelines. Numerous references have been cited in Chapter 8, related to rheological properties of waxy crudes. Wax particles in oil and water emulsions are known to affect rheology. Thus, the influence of wax flushed into a pipeline system cannot be ignored; similarly, the removal of wax from the flow due to solids deposition must be considered over large distances. Both of these effects influence the interplay between deposition and erosion discussed in Chapter 8. Note that the yield stress of wax adhering to pipeline walls depends on many factors, e.g., aging, history, encapsulated oils, shear-rate-dependent porosity, impurities, and so on.

The baseline rheological properties of oil and water emulsions are important to pipeline flows. Particularly useful is "Chapter 4, Rheology of Emulsions," in *Emulsions: Fundamentals and Applications in the Petroleum Industry* (Schramm, 1992). This publication explains the rheological classification of fluids, discusses numerous constitutive models, and describes various measurement instruments. The viscosity of an emulsion, the author notes, depends on (1) the viscosities of continuous and dispersed phases, (2) the volume fraction of the dispersed phase, (3) the average particle size and particle size distribution, (4) the shear rate, (5) the background temperature, and (6) the nature and concentration of the emulsifying agent. This reference also gives "theoretical" emulsion viscosity equations, plus numerous empirical formulas, considering the effects of added solids, their sizes, shapes, and distribution.

Other questions arise in subsea applications. For example, natural gas hydrates represent a potentially large source of hydrocarbons. They can be transported by mixing in ground form with refrigerated crude to create "hydrate slurries." How finely should the crystals be made? Obviously, the "solids to liquid" fraction will affect the rheology, which in turn affects pipeline economics. This interesting area of research will shed light on the practicality of hydrates as a viable energy source. Another problem area of interest is "flow start-up," since large pressure gradients will be needed to initiate movement once "gel" forms after flow stoppage. It is not clear that steady-state analysis methods will alone suffice. There is evidence that gels are "viscoelastic," that is, "memory" and time-history may figure into the flow equations. Additional rheological analysis of gels formed from waxy oils is desirable.

SOFTWARE NOTES

All of the simulators described in this book, with the exception of certain proprietary solids deposition models, are available for general use. The original MS-DOS based "Petrocalc 14" program offered in 1991, for annular flows only, gave but simple textual output for computed quantities. The completely revised program is now Windows compatible, supports both annular and duct flows, and supplements ASCII character plots with sophisticated color graphics. Also, the convergence acceleration methods described in this chapter have been fully implemented, and typical simulations (for grid generation and flow solution combined) now require only *seconds* on Pentium class computers. Integrated "text-to-speech" output is implemented in the software. In addition, prototype versions of our "fast Laplace equation solvers" have been developed, and are presently operable. Users and researchers interested in developing rheological models and partial differential equation solvers further, and in collaborating in future work, are encouraged to contact the author directly through his email address, "wilsonchin@aol.com." Portions of the computational mapping research reported in this book were supported by the United States Department of Energy under Grant No. DE-FG03-99ER82895.

REFERENCES

Bern, P.A., van Oort, E., Neustadt, B., Ebeltoft, H., Zurdo, C., Zamora, M., and Slater, K.S., "Barite Sag: Measurement, Modeling, and Management," *SPE Drilling and Completion Journal*, Vol. 15, March 2000, pp. 25-30.

Chin, W.C., "Irregular Grid Generation and Rapid 3D Color Display Algorithm," Final Technical Report, United States Department of Energy, DOE Grant No. DE-FG03-99ER82895, May 2000.

Gray, G.R., and Darley, H.C.H., *Composition and Properties of Oil Well Drilling Fluids*, Gulf Publishing, Houston, 1980.

Harris, P.C., and Heath, S.J., "Rheological Properties of Low-Gel-Loading Borate Fracture Gels," *SPE Production and Facilities Journal*, November 1998, pp. 230-235.

Huilgol, R.R., and Phan-Thien, N., *Fluid Mechanics of Viscoelasticity*, Elsevier Science B.V., Amsterdam, 1997.

Schramm, L.L., *Emulsions: Fundamentals and Applications in the Petroleum Industry*, American Chemical Society, Washington, DC, 1992.

Tamamidis, P., and Assanis, D.N., "Generation of Orthogonal Grids with Control of Spacing," *Journal of Computational Physics*, Vol. 94, 1991, pp. 437-453.

Thompson, J.F., "Grid Generation Techniques in Computational Fluid Dynamics," *AIAA Journal*, November 1984, pp. 1505-1523.

Thompson, J.F., Warsi, Z.U.A., and Mastin, C.W., *Numerical Grid Generation*, Elsevier Science Publishing, Amsterdam, 1985.

Index

A

Apparent viscosity, 27, 28, 70, 71,
  78, 116, 124, 134
Approximate methods, 11, 13, 25,
  76

B

Barite sag, 101, 114
Bingham plastic, 8, 27, 52, 59, 188
Bipolar coordinates, 14
Borehole stability, 18
Branch cut, 34, 35, 244, 245
Bundled pipeline, 2, 6, 159, 163,
  164, 166

C

Cementing, 18, 135
Clogged pipe, 2, 153, 186, 222
Cohesion, 195
Coiled tubing, 18, 153
Curvilinear grid, 33, 180
Cuttings transport, 17, 43, 65, 110,
  115, 119, 120, 125, 129

D

Dean number, 240
Deformation tensor, 26, 27, 31, 70
Density stratification, 100, 102,
  110, 114, 239
Dissipation function, 32, 85
Drilling applications, 6, 250
Duct flow, 17, 19, 168, 180

E

Ellis fluid, 10
Emulsion, 122, 251
Erosion, 194, 204

F

Finite difference, 4, 20, 35, 170,
  177, 182, 246
Foam, 15, 250

H

Hagen-Poiseuille, 1, 8, 170, 233
Heat generation, 18
Herschel-Bulkley, 9, 27, 218
Hydrate plug, 2, 6, 17, 19, 200,
  210, 213, 215, 218

M

Mudcake buildup, 191
Multiply-connected, 3, 243, 245

N

Navier-Stokes, 1, 16
Newtonian flow, 1, 7, 10, 16, 20,
  45, 54, 72, 74, 81, 168, 169,
  174, 210, 213
Non-Newtonian, 3, 26

P

Particulate settling, 10
Pipe bends, 231, 234, 236, 237
Plug flow, 5, 36, 219
Plug radius, 8, 9, 28
Power law fluid, 8, 27, 57, 61, 81,
  215

R

Recirculating vortex, 11, 16, 100,
  101, 239
Rectangular duct, 174, 177, 184,
  185
Rheology, 1, 5
Rotating flow, 15, 130, 146, 150

S

Scalability, 206
Secondary flow, 231, 238
Sedimentary transport, 195
Shear rate, 7
Singly-connected, 242
Slurry transport, 196
Solids deposition, 190, 204
Spotting fluid, 18, 131
Square drill collar, 50
Start-up conditions, 17
Streamfunction, 103
Stuck pipe, 18, 100, 110, 131, 186

T

Temperature effects, 15, 163, 164
Thompson mapping, 33, 182, 237,
  242, 246
Tridiagonal matrix, 172

V

Viscous stress, 26, 32, 70, 82, 116,
  119

W

Water-base muds, 115
Wax buildup, 2, 6, 17, 197, 198,
  203, 205

Author Biography

Wilson C. Chin earned his Ph.D. at the Massachusetts Institute of Technology, and his M.Sc. at the California Institute of Technology, in aerospace engineering, applied math, and plasma physics. His interests include fluid-dynamics, computational modeling, and also, electromagnetic simulation. He has written over fifty journal articles in aeronautical and petroleum engineering, and holds over twenty United States and international patents in formation evaluation, signal processing, mechanical design, and Measurement-While-Drilling.

Mr. Chin also authored four earlier books with Gulf Publishing Company, namely, (i) *Borehole Flow Modeling in Horizontal, Deviated, and Vertical Wells*, (ii) *Modern Reservoir Flow and Well Transient Analysis*, (iii) *Wave Propagation in Petroleum Engineering*, and (iv) *Formation Invasion, with Applications to Measurement-While-Drilling, Time Lapse Analysis, and Formation Damage*. He is presently working on a sixth, tentatively entitled *Borehole Electrodynamics*, which will apply state-of-the-art computational methods to well logging and resistivity earth imaging.

Prior to forming StrataMagnetic Software, LLC, Houston, Texas, in 1999, a research organization focusing on mathematical modeling and software development, Mr. Chin worked with Schlumberger, British Petroleum, and Halliburton Energy Services, building upon his early professional experience acquired at the aerospace leader Boeing Commercial Airplane Company. A perpetual optimist, Mr. Chin foresees strong industry demand for technology in the new millennium. As proof, he asks, "Who'd ever think 'hole-cleaning' and 'clogged pipes' would lead to computational rheology?" Certainly, not academic researchers; miracles like this can only happen in the Oil Patch.

Printed and bound by CPI Group (UK) Ltd, Croydon, CR0 4YY

08/05/2025

01864838-0002